D1062168

DIRECTIONS IN DEVELOPMENT

Livestock Development

Implications for Rural Poverty, the Environment, and Global Food Security

Livestock Development

Implications for Rural Poverty, the Environment, and Global Food Security

DIRECTIONS IN DEVELOPMENT

Livestock Development

Implications for Rural Poverty, the Environment, and Global Food Security

Cornelis de Haan
Tjaart Schillhorn van Veen
Brian Brandenburg
Jérôme Gauthier
François Le Gall
Robin Mearns
Michel Siméon

The World Bank
Washington, D.C.

1818 H Street, N.W.
Washington, D.C. 20433, USA

Manufactured in the United States of America
First printing November 2001
1 2 3 4 04 03 02 01

Cover photo credit: Food and Agriculture Organization of the United Nations; Peru.
Cover background: Curt Carnemark, undated, Nigeria.

ISBN 0-8213-4988-0

Library of Congress Cataloging-in-Publication Data has been applied for.

Contents

Preface vii

Acronyms and Abbreviations ix

Executive Summary xi
Action Plan xiii

1. Driving Forces 1
Driving Force 1: Growing Demand and Structural Change in the Livestock Sector 1
Driving Force 2: Changing Macroeconomic and Institutional Environments 4
Driving Force 3: Changing Functions of Livestock in the Developing World 6

2. Challenges 9
Food Security 9
Environment 16
The Poor 20
Livestock and Consumers 21
Animal Welfare 22
Implications for the World Bank 23

3. Main Interventions 25
Ensuring an Appropriate Policy Framework 25
Ensuring Access to Pastoral Land 30
Ensuring Access to Knowledge 32
Ensuring Access to Financial Services 43

Ensuring Access to Animal Health Services 46
Ensuring Access to Breeding Services 52
Ensuring Access to Markets 54

4. The Bank's Livestock Portfolio: Past, Current, and Future 61
The Future 63

References 69

Boxes
2.1 Targeting the Poor with Animal Products 15
2.2 Vertical Integration and Smallholders: The Example of the Turkish Development Foundation 21
3.1 Operation Flood: How a Commodity Project Can Reduce Poverty 30
3.2 Mongolia and Publicly Owned Pasture 31
3.3 Biotechnology and Livestock 35
3.4 Contracting Livestock Extension to Private Veterinarians: The Case in Mali 38
3.5 Nomadic Education: A Special Challenge for a Pro-People Livestock Strategy 42
3.6 Asian Development Bank Bangladesh Participatory Livestock Project 44
3.7 In-Kind Credit on Java 45
3.8 Livestock Services and the Poor: The Case in India 50
3.9 Drought Subsidies to Livestock Traders in the Isiolo District, Kenya (1996) 57
4.1 Livestock, Environment, and Development 64

Figures
4.1 Average Annual Lending (Total Project Costs) for Livestock-Only and Livestock Component Project Costs Funded by the World Bank since 1974 61
4.2 Percentage of the World Bank's Fiscal Year 2001 Livestock Portfolio (US$1.5 billion) as Allocated by Region, Species, and Activity 62

Tables
1.1 Sample Household Surveys on the Importance of Livestock in Rural Livelihoods 7
2.1 Actual and Projected Annual Growth Rates for Meat and Milk 10
2.2 Feed Conversion for Main Species and World Regions 11
3.1 Classification of Animal Health Services by Public or Private Sector Activity 48

Preface

In two decades, a significantly changed livestock sector is projected to produce about 30 percent of the value of global agricultural output and directly or indirectly use 80 percent of the world's agricultural land surface.[1] It will thus become the world's most important agricultural subsector in terms of value added and land use. The accelerated growth of livestock production and processing will require far-reaching changes in the roles of the public and private sectors in livestock development. This, in turn, warrants a reassessment of the role of international funding agencies that support livestock development.

The global livestock sector is changing fast. With a strong and growing demand, rapid institutional and macroeconomic policy changes, and a fundamental shift in the functions of livestock, there is a significant danger that the poor are being crowded out, the environment eroded, and global food security and safety compromised.

Livestock Development: Implications for Rural Poverty, the Environment, and Global Food Security was prepared by a group of livestock specialists from the World Bank with input from its partners. In this book, the Animal Resources Team (ART) argues that livestock can play an important role in poverty reduction, that the effects of livestock on the environment can be adequately managed, and that livestock can make an important contribution to global food security. This will only happen, however, if an appropriate policy framework is put in place, one that

1. Based on total current global agricultural land area of 4.9 million square kilometers, of which 3.4 million square kilometers are in grassland and 0.3 million square kilometers are used for grain for livestock feed (FAO 2001). Grassland will remain stable, and the area under feed grain is projected to increase to 0.5 million square kilometers.

enables the introduction of equitable, clean, and safe technologies throughout the food chain. The promotion of these enabling environments—particularly in areas where there are considerable market failures, such as equity, environment, and food safety—is seen as a core function of the public sector, including international financial institutions such as the World Bank.

This report provides recommendations on how to better manage ongoing changes in livestock development. First, it presents an overview of the main trends that can be expected to drive the sector over the next decades. Second, it discusses the negative or positive social, environmental, and health repercussions of those trends, and the institutional, policy, and technical requirements necessary to manage them. It concludes with a section on the current Bank portfolio in the livestock sector and provides an action plan for the future. The book generally describes good practices where available and identifies strategy implications of past experience.

Finally, this report summarizes the current thinking of the Animal Resources Team on the type of activities that the World Bank should support. It is a direct input into the review of the Bank's rural strategy. The book was written on the basis of a number of preparatory studies. Some were commissioned by ART and carried out with the explicit purpose of being introduced here—such as the work on food safety and microcredit for livestock, processing, and livestock extension services. Other studies were carried out in close association with ART staff—such as most of the work on the environmental and social aspects of intensification. Three important building blocks for this book are the work on "Livestock and the Environment: Finding a Balance" (de Haan and others 1997), "Livestock to 2020: The Next Food Revolution" (Delgado and others 1999), and "Contribution of Animal Agriculture to Meeting Global Human Food Demand" (CAST 1999).

Acronyms and Abbreviations

CAS	Country assistance strategy
CAST	Council for Agricultural Science and Technology
CBPP	Contagious bovine pleuropneumonia
CFAF	Communauté Financière Africaine franc
CIPAV	Centro de Investigación de Agricultura del Valle
FAO	Food and Agriculture Organization of the United Nations
GEF	Global Environment Facility
ICARDA	International Center for Agricultural Research
IMPACT	International Model for Policy Analysis of Agricultural Commodities and Trade
LEAD	Livestock, Environment, and Development Initiative
LID	Livestock in Development
NGO	Nongovernmental organization
OECD	Organisation for Economic Co-operation and Development
PKSF	Palli Karma-Sahayak Foundation
PRSP	Poverty reduction strategy paper
RAC	Regional agriculture chamber
Rs	Rupees
TKV	Turkish Development Foundation

Executive Summary

Demand and supply patterns for meat and milk, the roles of livestock, and the external economic and institutional environment in which the livestock sector operates are rapidly changing at the global level. These changes require a major adjustment in the role of international agencies such as the World Bank in supporting the contributions livestock can make to poverty reduction, environmental sustainability, and food security. First, in the next two decades the livestock sector is projected to become the world's most important agricultural subsector in terms of value added and land use. The growing, increasingly urban, and more affluent population in the developing world will most likely demand a richer, more diverse diet, with more meat and milk products. As a result, global meat demand is projected to grow from 209 million tons in 1997 to 327 million tons in 2020, and global milk consumption from 422 million tons to 648 million tons over the same period. This is appropriately called the "livestock revolution" (Delgado and others 1999).

The developing world is projected to be the most important supplier to this growing market. Production of meat and milk is expected to increase by about 3 percent per year in the developing world, compared to about 0.5 percent in the industrial countries. Industrial poultry production could be the fastest growing sector with an expected increase in output of about 80 percent until 2020. The other livestock commodities will grow at about 50 percent over that period. Stricter environmental regulations; consumer concerns about health and animal welfare; increases in the price of grain, water, energy, and transport; land scarcities; and major breakthroughs in the use of tropical fodder might shift the balance back to red meat production. Whatever happens, the livestock sector will undergo some dramatic structural changes.

The traditional roles of livestock in the developing world are changing, although countervailing trends are also emerging. Tractors are replacing animal traction in Asia, although cattle are still an important and growing power source in African agriculture, and horse-drawn agriculture is making a temporary comeback in the former Soviet Union. Inorganic fertilizer is replacing manure, although emerging consumer preferences for organically produced food might again increase the demand. There will be a growing divergence between industrial production systems and smallholders. New and better rural microfinance systems are substituting for the role of banking in livestock, although in some other regions livestock have become the main social safety net for retirees. Finally, livestock are still an important component of the livelihood of about 750 million rural poor and can be a critical input in the dominant starch diet of those poor, reducing malnutrition and improving their lives.

The institutional context in which the sector operates is changing. Over the last decade, public and private sector roles have become increasingly stratified, with an emphasis on government doing better with less. Decisionmaking is slowly becoming more decentralized, and new, more programmatic but less directed investment instruments are being introduced. Finally, in most middle-income countries, the industry is becoming concentrated in large commercial units, with improved short-term efficiency but also greater vulnerability to disasters, whether epizootic disease, environmental calamities, or mishaps related to food safety.

This document argues that under any scenario the increase in demand will put strong pressure on global natural resources. This increase might crowd out the poor, endanger global food security and food safety, and affect animal welfare. The environmental effects include land degradation in marginal arid lands, erosion of biodiversity in tropical savannas and rain forests, and nutrient loading and water pollution by industrial producers around the main urban areas of the developing world. Equity effects include crowding out smallholders from the sector because economies of scale for large units do not adequately internalize environmental costs and increased encroachment of communal areas for ranching and producing feed grain. The price of grain may increase and also affect the urban poor. The shift to more grain-based production could seriously affect global and national food security, whereas the overconsumption of animal products by the middle-income classes in the developing world might lead to diet-related chronic disease patterns similar to those in the industrial world. The increased concentration of livestock might indeed lead to an increase in the emergence of new disease patterns and the incidence of food-borne diseases. Concentrating production may also lead to practices unfriendly to animal welfare, such as poultry batteries and sow crates.

Action Plan

The livestock portfolio analysis shows that our current World Bank operations still lack a specific poverty and environmental focus, an approach this book advocates. This lack of focus is shown by the low level of investment in the poorest regions of the world (central Asia, South Asia, and Sub-Saharan Africa) in pastoral development and small stock and, to some extent, in the low share of investments to improve animal health and nutrition, which are the critical constraints faced by the poor.

In this context, this book argues for a World Bank livestock strategy with a people-focused approach, giving high priority to the public goods aspects of poverty reduction, environmental sustainability, food security and safety, and animal welfare.

For a stronger focus that favors the poor, the following actions are required:

a. Ensure that in analytical (economic and sector work), strategic (poverty reduction strategy papers and country assistance strategies [CAS]), and project and program documents, adequate attention is paid to the institutional, incentive, and technology framework of those production systems that are practiced by the rural poor. For small-scale mixed farmers, the policy dialogue and operations would apply past lessons in livestock service delivery and integral food chain management. This application would assess the incentive framework for smallholders (eliminating eventual bias for industrial production); cover all functions of livestock; pay increased attention to producer organizations in input supply; include processing, marketing, and participation in the policy dialogue; support continuing privatization of animal health, breeding services, and multiple-source advisory services; and refocus research, extension, and education on the needs of the poor.

   For pastoralists or mountain herders, the program would increase support for pastoral development through continued piloting and upscaling along the lines of new thinking in range ecology, paying particular attention to pastoral empowerment, mobility and flexibility in access to resources and drought preparedness, including markets and appropriate financial instruments (insurance, savings) to mitigate drought and other risks. For those who practice diversification, production systems would include typical systems used by the poor (bees, rabbits, dairy goats, poultry, and so forth) and their relevant support systems. Finally, poor non-livestock keepers would be a potential beneficiary of livestock development.

b. In tandem with other agencies, develop a deeper understanding of the key aspects of pro-poor design of livestock development operations, training modules, and enhanced awareness in pro-poor livestock development strategies for decisionmakers involved in the Poverty Reduction Strategy Paper (PRSP) and rural strategy formulation.
c. Large-scale commercial, grain-fed feedlot systems and industrial milk, pork, and poultry production would not be financed by the Bank except for the public good aspects of environment and food safety.

For a stronger focus on mitigating the negative and enhancing the positive effects of livestock development on the environment, the following specific actions need to be taken:

a. Integrate livestock-environment interactions into environmental impact assessments and national environmental action plans. Some of the key issues would include incentive and regulatory distortions favoring large producers, internalization of environmental costs in the price, and common resource access.
b. Continue developing innovative approaches to managing the interactions between livestock production and the environment in the "hot spots." Such approaches include drought preparedness to address desertification of arid rangelands, benefit-sharing systems for livestock-wildlife systems, payment for ecological services in degraded pastures to reduce deforestation in the humid tropics, and areawide integration of industrial units to limit nutrient loading and groundwater pollution. Close cooperation with the Livestock, Environment, and Development (LEAD) Initiative and the Global Environment Facility (GEF) could help in testing and broadening current experience.
c. Mainstream sound ecological farming practices such as integration of crops and livestock, development of markets for organic products, and so forth.

For a stronger focus on food safety and health issues, programs must at least concentrate on the following three areas:

a. Policies and institutions related to the level of involvement in food safety for domestic consumption and export, control of diseases of trade, emerging diseases and their effect on human health, and, consequently, strengthening links with the health sector (the CAS, the PRSP, and multisector projects).

b. Appropriate legislation, if needed, adapted to local food preparation practices and trends, the role of the public sector in food safety, and partnerships with the private sector and consumers.
c. Infrastructure and human and institutional capacity building in general but particularly in sustainable animal health and production, best practices in managing the food chain from farm to fork, and informed participation in organizations for setting international standards, so that the voices of the developing countries are heard.

Institutionally, the Bank should direct its human and other resources toward this agenda, particularly the following:

a. Continue to provide leadership in international livestock-related initiatives such as LEAD, livestock and poverty, sustainable dry land management, and food safety. Leadership should also be provided to the international community on how to integrate such pro-poor livestock development design into the policy dialogue (PRSP).
b. Focus on strategic staffing, using the considerable attrition over the next years to attract staff who can respond to the new focus on poverty reduction, environmental sustainability, and food safety. This will require several actions. First, maintain a livestock development adviser at the Rural Development Department with excellent development and private sector credentials who can further develop the international discussion on critical current and emerging global issues in livestock development. The adviser would play a key role in promoting livestock development for poverty alleviation by advocating the integration of livestock development in the PRSP, the CAS, and so on and maintaining high-quality standards for livestock development in Bank operations.

   Second, in the regions with strong pro-poor livestock development needs (Sub-Saharan Africa, Central Asia, and South Asia), maintain, or appoint, high-level livestock generalists.[1] These generalists could be the leaders who shape policy dialogue on pro-poor, sustainable livestock development in their respective regions in sector work and projects and programs, prepare and implement such operations, expand the development dialogue to other sectors (health, private sector development, and so forth), and

1. This is especially relevant in those regions (Africa, Europe, and Central Asia) where there has been an erosion of national technical and livestock economic skills over the last few decades.

work in cooperation with other international organizations such as the Food and Agriculture Organization of the United Nations (FAO)/Cooperative Program.

Third, develop closer cooperation with the International Finance Corporation and, together, show how commercial livestock development can be used for poverty alleviation. Fourth, maintain a formal or informal network of livestock specialists.

With such an action plan, the Bank's contribution to addressing the environmental, social, and health issues resulting from one of the most significant structural changes in the agricultural sector will be greatly enhanced.

1

Driving Forces

The growing demand for meat and milk in the developing world, changing functions of livestock, changing international and national socio-economic policy frameworks, and changing consumer perspectives are likely to be major driving forces in the global livestock sector during the next two decades.

Driving Force 1
Growing Demand and Structural Change in the Livestock Sector

The most likely trends for the demand and supply of meat and milk, and the structural shifts of the industry at the global level, are quite startling.[1]

While there might be countervailing forces (see below), by 2020 the global population is projected to consume about 120 million tons of meat and 220 million tons of milk above current consumption. To put this in perspective, a similar increase in demand occurred over the last 40 years. The global population of 8,000 million in 2020, which will be increasingly urban and more affluent, will demand a richer and more diverse diet with an increasing share of meat, milk, and eggs. As a result, total global meat demand is expected to grow from 209 million tons in 1997 to 327 million tons in 2020 (56 percent). Over the same period global milk consumption is expected to increase from about 422 million tons to 648 million tons (54 percent). Delgado and others (1999) and Delgado, Rosegrant, and Meyer (2001) call this—appropriately—the "livestock revolution."

1. This section is mostly based on Delgado and others (1999) and Delgado, Rosegrant, and Meyer (2001).

Most of the growth in demand will be in the developing world because, for the lower income classes, meat and milk have high income elasticity. For example, in countries with per capita annual incomes between US$1,000 and US$10,000, Schroeder, Berkeley, and Schroeder (1995) found that the income elasticity for meat varied between 1 and 3 depending on the type of meat. Above US$10,000 income, income elasticity levels are up to 1. As a result, from 1997 to 2020 Delgado, Rosegrant, and Meyer (2001) estimated that per capita meat consumption in the developing world will increase from 25 kilograms to 35 kilograms, compared to an increase of 75 kilograms to 84 kilograms in the industrial world. Comparable figures for milk consumption show an increase of 43 kilograms to 61 kilograms per capita in the developing world and of 194 kilograms to 203 kilograms in the industrial world. This indicates that 80 percent of the growth in total demand for meat and 95 percent of the growth in total demand for milk are expected to occur in the developing world. It also implies that the developing countries' share of total global meat consumption will increase from the current 53 percent to about 65 percent in 2020. The comparable figures for milk would be 43 to 57 percent. This is another critical characteristic of the livestock revolution.

Similarly, most of the growth in production is projected to occur in the developing world. Over the last decades, the percentage of meat and meat products that has been traded internationally has remained a stable 14 percent of total global consumption (McCalla and de Haan 1998). Globalization might increase trade, but infrastructure constraints (port facilities) in the developing world, higher transportation costs, and stricter animal welfare and environmental regulations in the industrial world may support a shift toward increased production in the developing world. The projections of Delgado, Rosegrant, and Meyer (2001) confirm this with estimates that the production of meat in the developing world is expected to increase from 110 million tons in 1997 to 206 million tons in 2020, and milk from about 208 million tons in 1997 to 386 million tons in 2020. According to these latest projections, by 2020 63 percent of most global meat and slightly more than one-half (50.3 percent) of milk are expected to be produced in the developing world.

Livestock production is expected to shift from temperate and dry regions to more humid and warmer regions. A clear worldwide shift from the temperate regions has already occurred. For example, in the United States production has moved to the southern states, and in the South American tropics from temperate highlands to subhumid savannas. In Brazil, the share of cattle in the subhumid *cerrados* has risen from 14 percent of the national population in the 1940s to 29 percent by 1990 (Toledo 1990). A similar trend is occurring in Africa, with strong increases of livestock numbers in the subhumid savannas (Bourne and Wint 1994).

Poultry will be the main source of growth, with other sectors showing similar growth patterns at a lower level. Poultry have a better feed conversion ratio than pigs and ruminant animals, and their production technology is more universal. For these and other reasons, worldwide poultry production is expected to increase by almost 80 percent over the period 1997–2020, whereas dairy, beef, and pork production are projected to increase by 40–50 percent over the same period.

Finally, an increasing share of the production is projected to come from industrial forms of production. Economies of scale, increasing labor costs, and declining capital costs will promote further industrialization. Over the period 1983–93, industrial meat production grew at an annual rate of 4.3 percent, twice the 2.4 percent growth in mixed farming, and six times the growth rate of 0.7 percent per year for meat production from grazing systems (de Haan, Steinfeld, and Blackburn 1997). This industrialization of production and processing is expected to continue over the coming decades.

Countervailing Forces

There could, however, be countervailing forces to these trends. First, the comparative advantages of economies of scale of industrial pig and poultry production might disappear if the "polluter pays" principle is invoked in the developing world and the environmental costs of excess nutrient emissions are applied.

Second, the recent outbreaks of pandemics such as classical swine fever or foot-and-mouth disease have focused consumer attention on the negative sides of intensive production. In Europe this is leading to reduced consumption and proactive policies to promote more extensive production methods, although the persistence of these trends is unclear.

Third, because of the shift to grain-based pig and poultry production, the large increase in global feed requirements could increase grain prices and thus reverse the balance from grain feeding to grass-based systems, or lower consumption levels. The economic International Model for Policy Analysis of Agricultural Commodities and Trade (IMPACT) used by Delgado, Rosegrant, and Meyer (2001) does not account for eventual resource constraints such as water, land, and significant increases in energy costs.

Fourth, intensive systems require more energy per kilogram of meat than the more extensive land-based systems, mainly because of the high energy and water requirements for feed production. Increases in the price of energy would shift the balance back to grass-based systems.

Fifth, a major breakthrough in the production of high-quality fodder in the tropics or improved digestibility of the current high-fiber tropical forages could radically shift the balance from pigs and poultry to cattle and small ruminants, and from industrial production to grazing systems. It would also shift production to subhumid tropical areas, with their potential for high levels of biomass production.

DRIVING FORCE 2
Changing Macroeconomic and Institutional Environments

The growth in livestock production will have to take place in greatly changed macroeconomic and institutional environments.

Key Changes

A focus on poverty reduction and increasing privatization are among the changes that may affect livestock development, particularly the role of public institutions in this area.

External development funding will increasingly be directed to poverty reduction. The greatly sharpened focus of international agencies on poverty reduction is probably the most important institutional change in recent development thinking. All international development agencies now have selected poverty reduction as their main focus. With about 70 percent of the poor living in rural areas (World Bank 1997), this should lead to a higher priority for investments in rural development. Livestock play an important role in rural poverty reduction (chapter 3), but what is needed is a much greater understanding of the profile of poor livestock keepers, the enabling environment for equitable growth of the sector, and the institutions and technologies poor livestock keepers require.

The distribution of responsibilities among the public and private sectors will continue to evolve. Over the last decade, the distinction between public and private sector roles in livestock development support activities has become increasingly focused. Economic adjustment policies, financial crises, and the disappointing performance of semipublic and public sector services have led public funding to be much more strictly directed to public goods. International public funding for commercial activities is practically abolished, and most public funds are now allocated to the funding of goods and services with clear externalities such as free riders, moral hazards, or other forms of market failure. Over the last decade, budgets of livestock services have been cut, at least in real terms, in many countries. In a recent survey of 16 countries in Sub-Saharan Africa, a decline in the number of civil servants and a reduction in the share of salaries in the overall budget was reported (Gauthier, Siméon, and de Haan 1999). The same study found a decrease of 11 percent in the staff of veterinary services over the last 10 years. This has resulted from increasing commercialization and privatization of services (Umali, Feder, and de Haan 1992).

Global trade in livestock commodities is expected to increase. Approximately 150 million tons, or about one-third of internationally traded agricultural commodities, are livestock products or livestock feed. The increase in demand, in addition to declining levels of protection and export subsidies ("dumping") under the World Trade Agreement, may open new opportunities for developing countries, but it also puts increasing demands on their animal health and food safety standards. Already the value of livestock products exported by developing countries almost equals that of cereals. It is being increasingly realized that domestic demand is not sufficient as an engine of growth for rural areas, and that exports are essential for robust rural income growth. Opening export markets for livestock products could be such a growth engine for the rural poor.

Decisionmaking will increasingly be decentralized. Increasing involvement of local communities in decisionmaking is a worldwide trend. This affects livestock development directly because it often concerns decisions dealing with natural resource management (access to land and water, environmental issues), and the management of public services (animal health and extension, and so forth). Decentralized services and producer organizations now form a central part of the development strategies of many African, Latin American, and South and Central Asian countries.

The structure of the industry will continue to change. Vertical integration linking input suppliers (feed, stock), producers, processors, and supermarkets is already the common structure in the member countries of the Organisation for Economic Co-operation and Development (OECD). With increasing industrialization of livestock production, vertical integration of the sector is also becoming more common in Latin America and East Asia. In Thailand, for example, 80 percent of poultry production comes from only 10 large, vertically integrated companies that supply under contract feed and day-old chicks to medium- and large-size producers (Henry and Rothwell 1996). Similarly, in Brazil, four integrators cover about 40 percent of the broiler market.

The type of investment instruments in development operations will change quite dramatically. Already, support for livestock development is increasingly integrated as components of broader development programs rather than as stand-alone operations. For example, the number of livestock-only projects approved each year by the World Bank has declined from about ten in the 1970s to approximately one today. The number of livestock components has remained much more stable (chapter 4), but the shift toward more comprehensive development support continues. PRSPs that lead to multisector projects, program lending, and comprehensive poverty reduction credits will gain more importance over the coming years. The integration of livestock as a component within broader rural development operations, while conceptually sound, could lead to a deterioration of the technical quality of livestock component activities as declining budgets for preparation and supervision reduce the resources available for technical inputs. This will become more important if even broader investment instruments are used. Ensuring that rural areas, and the poor livestock keepers in those areas, are technically and financially covered adequately will be one of the greatest challenges for the near future (see chapter 4).

DRIVING FORCE 3
Changing Functions of Livestock in the Developing World

Structural changes in global agriculture will also cause the livestock sector to change from a multifunctional to a commodity sector.

Main Functions of Livestock

Although livestock are sometimes associated with the rich, they can be a critical input to help poor rural people escape the poverty trap. For more than 600 million poor people, livestock are an important component of household income (LID 1998). For others, livestock could also be an important component of household income. In many poor areas of the world, particularly where land is collectively owned, livestock owners might be wealthier in assets than crop farmers, but the income from these assets can be substantially lower. For example, income and most other social indicators of pastoralists in West Africa are substantially lower than those of arable farmers, although these pastoralists are wealthier in absolute assets.

Livestock raising also carries much more risk than crop production. It takes longer to restore its productive capacity after a drought because rebuilding the herd requires several years, whereas crop production can be restored after one good year. Delgado and others (1999) present a series of field studies that generally show that livestock provide a higher share of the income of the rural poor than wealthier farmers (table 1.1).

As a source of income, milk and meat production provide approximately 26 percent of the agricultural gross domestic product in the developing world (compared to 55 percent in the industrial world). Livestock are still the source of power to till one-half of the developing world's arable land, and produce about 15 percent of the world's crop nutrients (de Haan, Steinfeld, and Blackburn 1997). Finally, the demand for natural fiber is still high and even increasing in some places.

Table 1.1. Sample Household Surveys on the Importance of Livestock in Rural Livelihoods

Country	*Wealth or poverty indicator*	*Stratum*	*Percentage of population*
Ethiopia	Household income	Very poor	6
		Poor	24
Egypt	Land holdings	Landless	63
		Largest landholding	14
Kenya	Household income from dairy business	Lowest 1/5	61
		Highest 1/5	38
Pakistan	Land holdings	Landless	14
		Largest landholding	11
Pakistan	Household income	Lowest 1/5	25
		Highest 1/5	9
Philippines	Household income	Lowest 1/5	23
		Highest 1/5	10

Source: Delgado and others (1999).

As a capital asset, livestock form a key source of petty cash, paying for school fees, medical costs, and so forth. In many situations they are the sole instrument for savings and insurance, because more formal banks are unreliable, too remote, or provide unattractive returns in situations with high rates of inflation. Investments in livestock are also used to hedge against unexpected natural disasters such as disease outbreaks, droughts, and floods, including traditional hedging and safety net systems. The provision of livestock through inheritance, gifts, and so on is part of most rural societies. Livestock allow the poor to gain private benefits in nutrients and income from communal areas. Much of dryland agriculture in Africa and South Asia is based on the transfer of nutrients from common rangeland areas to individual area plots.

Changing Functions of Livestock

These functions, however, are changing, although strong countervailing forces are present.

The "banking" function is probably declining in most parts of the world, although it is increasing in others. Following the increasingly successful establishment of informal and formal rural finance institutions such as the Grameen Bank in Bangladesh, and the decline in inflation rates in many countries, the attractiveness of direct investments in livestock for savings and insurance might decrease somewhat in parts of Asia, although exact data are not available. Credit, food aid, and alternative sources of income have also reduced the reliance on livestock as a buffer against calamities in parts of Africa (Fafchamps, Udry, and Czukas 1998). New population groups, such as retirees in the former Soviet republics and increased interest in urban agriculture have created a new set of smallholders. On the contrary, new investors in Sub-Saharan Africa, the Middle East, and Latin America are increasingly using livestock for investment and savings. Indeed, it has been estimated that more than one-half the livestock population in the Sahel is now owned by absentee owners (Fafchamps, Udry, and Czukas 1998).

Inorganic fertilizer is gaining importance in the developing world, although organic farming is also increasing in importance. Use of inorganic fertilizer is becoming increasingly important, especially in Asia. According to the FAO (2001), inorganic fertilizer consumption in the developing world rose from 65 million tons to 85 million tons over the period 1990 to 1999. In China alone it more than doubled from 17 million tons in 1986 to 36 million tons in 1999. The easier-to-handle and apply inorganic fertilizer is replacing manure in many areas of the world. However, an increasing interest is apparent in the production of organic milk and meat in both the industrial and developing worlds. For example, spurred by a proactive subsidy policy, sales of organic milk in Denmark increased from a 1 percent market share in 1986 to a 15 percent share 10 years later (Danish Ministry of Food and Agriculture 1999). It can be expected that the increasingly affluent populations of Latin America and Asia will demand similar products. Such organic production methods are reviving the interest in use of manure as an organic fertilizer. In addition, increasing evidence, also from the tropics, shows that the use of inorganic fertilizer in combination with manure leads to more profitable and sustainable agriculture.

In some parts of the world, mechanization is taking over. Between 1990 and 1999, the total number of tractors in use in the developing world increased from 5 million to 7 million, strongly driven by the increase from 0.6 million to 1.4 million in India alone (FAO 2001). However, over that period the number of tractors in use in Africa increased by less than 10 percent, and animal traction is still growing in importance on that continent (FAO 2001). Moreover, horse-powered cultivation has greatly increased in many parts of the former Soviet Union, since fuel subsidies were phased out and the fragmentation of the state farms wiped out the economies of scale for mechanized farming.

In summary, environmentally sustainable and pro-poor investments in livestock development will have to consider a very dynamic environment to address the key challenges emerging from the above trends.

2

Challenges

Macroeconomic and institutional changes in the livestock sector, along with the strong projected increase in demand, could severely affect global food security, the natural resource base, and rural equity. Curbing demand for meat and milk, as advocated by some (Goodland 1997) is not a viable option, as shown by past failures to reduce the demand for other products and substances.

Food Security

One of the most critical issues in the debate on the livestock revolution is its eventual effect on household, national, and global food security. If food security is defined as the individual's access to enough food to maintain a healthy and active life, food security needs to be reviewed for (a) availability, (b) accessibility, and (c) nutritional adequacy. In the debate on global food security, the effect of the livestock revolution on global cereal production is critical.[1]

Availability

To meet the predicted demand for livestock products, high annual production growth rates are expected to continue, especially in the developing world (table 2.1). To put it into perspective, assuming a stagnant global livestock population until the year 2020, beef production per head (cattle, yak, and buffalo) in the developing world would have to double from the current level of 20 kilograms per head per year to 41 kilograms in 2020.

1. This section is based mostly on McCalla (1999).

Table 2.1. Actual and Projected Annual Growth Rates for Meat and Milk

Commodity	Annual growth (percent)	
	1982–93	*1997–2020*
Meat (global)	2.8	2.0
Meat (developing countries)	5.2	2.9
Milk (global)	1.0	1.5
Milk (developing countries)	3.2	2.7

Sources: Delgado and others (1999); Delgado, Rosegrant, and Meyer (2001).

This growth is substantial, although still far below the current level in the Western world at 87 kilograms per head per year. A similar situation exists for milk.

Are these growth rates achievable? On the basis of past performance, Delgado and others (1999) predict that they can be met, but the growth would be mainly cereal-based. Some possibilities for increased grazing exist in the subhumid savanna areas of Africa, Central Asia, and Latin America, but over the last years, urban and crop encroachment have taken away some of the best grazing areas in temperate areas of the world. The area under grass and rangeland has remained stable at approximately 3.3 million square kilometers, of which about 2.5 million square kilometers are in the tropics (de Haan, Steinfeld, and Blackburn 1997). Thus the incremental demand for feed grains to fuel the livestock revolution would be quite significant. Delgado, Rosegrant, and Meyer (2001) project that the global feed grain demand will rise from 660 million tons in the 1997 to 925 million tons in 2020, with demand in the developing world increasing over the same period from 235 million tons to 432 million tons per year. For China alone the demand for feed grain would increase from 91–111 million tons[2] in 1997 to 221 million tons in 2020.

The feasibility of producing those incremental feed grain needs, and the implications for global natural resources, lies at the heart of the debate on global food security. Most feasibility estimates using economic models indicate that the incremental feed grain can be produced. These models are based on past growth of the biological yield of cereals and continuing intensification and increasing irrigated area (Delgado and others 1999; Mitchell and Ingco 1993). Some of the ecological models (Brown and Kane 1994) take into account environmental degradation, increased competition for water, and limited expansion of arable land, which leads

2. The variation represents extrapolations of the different assumptions from the U.S. Department of Agriculture (FAO 1997; Simpson, Cheng, and Miyazaki 1994).

to a much more pessimistic view. Moreover, a mismatch exists between production areas and most consumer concentrations. McCalla (1999) argues that it will be an enormous challenge that is nevertheless possible, provided that

- Sustainable production systems that are capable of doubling output can be developed, especially in the subhumid tropics,
- Policies that do not discriminate against agriculture are implemented,
- Investments in agricultural research remain high on the agenda, and
- Distortions to free agricultural trade are removed.

This book assumes that future demand for feed grain will strongly increase in the developing world, and that in any case this demand will put significant pressure on global land and water resources. An increased focus on policies and technologies, which would reduce the demand for feed grain, is therefore recommended. For the livestock sector to satisfy the growing demand within resource constraints, the priority would probably not be an increase in biological yield per animal, but would need to focus more on the efficiency of input use, especially for feed grain and energy. The two most critical factors affecting the efficiency of input use are the choice of species and systems and input conversion. Their effects are explained below.

Choice of species and systems is important because different systems have different feed and energy efficiencies, with an increasing efficiency from aquaculture, via milk, broilers, and eggs, to pork and feedlot cattle.[3] Table 2.2 gives some ranges of total feed conversions and amount of edible grain used per kilogram of product.

Table 2.2. Feed Conversion for Main Species and World Regions

	Feed conversion		*Edible grain per kilogram of product*	
Species	*Kilograms feed per kilograms live weight gain*	*Kilograms feed per kilogram product*	*Industrial world*	*Developing world*
Aquaculture	1.2–1.6	1.5–2.0	n.a.	n.a.
Poultry meat	1.8–2.4	2.1–3.0	2.2	1.6
Pork	3.2–4.0	4.0–5.5	3.7	1.8
Beef	7	10	2.6	0.3

n.a. Not applicable.
Sources: CAST 1999; authors' estimate.

3. Based on decreasing feed conversion rates, that is, amount of feed required to produce 1 kilogram of live weight gain.

From a global perspective, increases in aquaculture and poultry production would put the least pressure on global food security. Some recent reviews show, however, that cattle production in the developing world is still mostly based on by-products that are not suitable for human consumption (bran, oil cake, and so forth), in addition to grass and forage. For example, the total global production of 54 million tons of human edible animal protein required an estimated 74 million tons of human edible plant protein, or a conversion of 1:1.4 (Steinfeld, de Haan, and Blackburn 1997). The Council for Agricultural Science and Technology (CAST 1999) calculates that the average grain consumption per 1 kilogram of beef in the OECD countries is 2.6 kilograms of edible plant food per 1 kilogram live weight gain.[4] In developing countries, only 0.3 kilograms of edible plant food is used to produce 1 kilogram live weight gain. Thus the use of human edible food for animal production currently concerns mostly industrial-country production systems, although the livestock revolution is causing these models to expand rapidly in the developing world. Such grain-based systems would generally not justify financial support, because of their negative resource implications, their commercial nature, and their potentially negative social effects.

From a food security and equity point of view, situations occur in which grain-fed ruminant production might be recommended and, therefore, warrants support

- As a buffer for grain stocks, especially but not exclusively, in areas with poor road infrastructure such as in the Sahel, where cattle fattening can be an excellent way to buffer fluctuations in grain production; and
- For smallholders, for whom it can be an effective way to alleviate poverty and provide some added value, such as work oxen.

Accessibility

The world still contains approximately 840 million undernourished people, and more than 1 billion people suffer from deficiencies in one

---

4. This contrasts with the feed conversion of 8 kilograms grain per 1 kilogram growth, or higher in cattle, often quoted in the literature (Pimentel 1997). However, concentrate feed in the so-called feedlot production is mainly used in the United States, and then only used for a small part of the life cycle because most of the growth (55–70 percent of the final weight) is produced from grazing on rangelands before the animal enters a feedlot. Moreover, while in the feedlot, the body composition of the animals changes, and the common use of kilogram of grain per kilogram body gain (rather than edible product) underestimates the conversion from feed grain into human edible food.

or more micronutrients. The critical factor leading to this high number of hungry people is probably not so much the physical availability of food at the household level, but the power to purchase food. The World Food Summit set a target to reduce the number of undernourished to 400 million by 2015. Meeting that objective will depend on price trends, purchasing power, and market infrastructure. The following estimates are available.

Food prices are likely to decline further. Even under the scenario described above of increased meat consumption, Delgado and others (1999) project a continuing declining trend in global grain prices, although less pronounced than has occurred over the last decades. The livestock revolution, therefore, would not affect access of the poor to basic food commodities. Prices of livestock products would also decline over the next two decades, although less than those of cereals, and except for small quantities, milk and meat would probably remain out of reach for the urban poor. The contribution of the livestock sector to poverty reduction, therefore, would be mostly for poor livestock producers, rather than for poor rural and urban nonlivestock-producing consumers.

The appropriate policies, institutions, and technologies will affect success. The role of livestock as an income-generating tool will depend on the success of the policies, institutions, and technologies that are available. The intensification and concentration of the industry over the last decades threatens to crowd out the poor. It will therefore depend to a large extent on the level and success of pro-poor policies, institutions, and technologies to determine how successful livestock development will be for future poverty alleviation (see chapter 3).

Continued development of market infrastructure is required. Extensive areas, in regions of Africa and Asia, for instance, still exist where physical access to market outlets are nonexistent or are blocked during parts of the year. In other regions, such as the former Soviet Union, market access is also hampered by over-regulation and consequent rent seeking and risk of corruption.

Nutritional Adequacy

Animal food products can play a role in addressing malnutrition. Mild to moderate protein-energy malnutrition is prevalent throughout the developing world, especially among children. Micronutrient deficiencies often coexist with public expenditure management. The key constraint is that plant-based diets do not provide high-quality energy, protein, and micronutrients.[5] Although some leafy vegetables contain high quantities of iron, zinc, and calcium, their bio-availability is much lower than in meat and milk. Similarly, milk is an important source of calcium. It is difficult for a child to even approach calcium requirements on a cereal-based diet.

Good evidence can now be presented that animal-based food could provide these nutrients. The most comprehensive study, the Human Nutritional Collaborative Research Support Study, was funded by the U.S. Agency for International Development in Kenya, Egypt, and Mexico. It investigated diet and growth and child development and pregnancy outcomes. These studies show that

- Consumption of animal foods during pregnancy is a predictor of infant growth;
- Animal foods were clearly associated with postweaning growth.

For example, studies in such desperate countries as Kenya, China, Jamaica, Mexico, Nicaragua, and Brazil all found that children who consumed cow's milk obtained significantly greater height. Livestock ownership was positively associated with child nutritional status provided the milk was consumed at home and general sanitation conditions were appropriate for use of cow's milk. Toddlers who ate little or no animal protein did not perform as well on cognitive tests as those who had animal protein in their diets.

Protein and energy malnutrition and micronutrient deficiencies are directly associated with the suppression of the body's defenses against diseases. This synergy between infection and malnutrition is the leading cause of death in children under five years of age in developing countries. Minimal amounts of animal foods, as shown above, can play a critical role in reducing child mortality.

When confronting chronic malnutrition and micronutrient deficiencies, supplementing traditional diets of maize and cassava with animal foods can make an important difference. Public programs to reach vulnerable groups are thus justified and would need to be part of livestock development projects. The key issue is targeting groups from the urban and rural poor to ensure they get access to expensive animal foods. Experience with targeted food programs is limited, but one of the most interesting experiences concerns targeting the poor for milk in Tunisia (box 2.1).

---

5. For example, to meet the average daily requirements for energy, iron, and zinc, a child would have to consume 2 kilograms of maize and beans each day, which is much more than is physically possible. The same amount would be available in 60 grams of meat.

Box 2.1. Targeting the Poor with Animal Products

The results of traditional targeted programs, such as direct assistance schemes, are generally mixed. Targeted programs entail implementation difficulties and heavy administrative costs and lead to rent seeking. To limit these constraints, in 1990 the Tunisian government established a self-targeted program. Self-targeting occurs when benefits are available to all, but, as in the Tunisian case, because of perceived differences in quality, the nonpoor elect not to participate.

Consumption habits were analyzed to identify products that the wealthier would perceive as inferior goods because of unattractive packaging or ingredients that would cause them not to purchase them. Pasteurized, reconstituted milk (the least preferred type of milk) packaged in cheaper cartons, such as half-liter plastic bags, cartons, and flimsy milk pouches, were less attractive to wealthy consumers. Data from the 1993 household expenditure survey indicated that subsidies of this type of milk were indeed well targeted to the poor, because among consumers they were nearly the exclusive purchasers of this type of milk.

Source: World Bank data.

Relationship between animal food consumption and chronic diseases. Chronic diseases resulting from high consumption of animal products represent the negative side of the criterion for nutritional adequacy. On the basis of an extensive survey in a large population in Boston, Massachusetts, Hu and Willett (1998) offered the following conclusions:

- Different products can have different effects on the same disease, and the same product can have different effects. Clear distinctions should be made between red meat, which when consumed in large amounts[6] probably contributed to an increased incidence of cardiovascular disease and colon cancer, while the negative effects of white meat, dairy products, and eggs was less clear.
- It is probable that in the developing world the increased intake of saturated fat, including that from animal foods, by more affluent

6. Current annual consumption of 21 kilograms of meat and 40 kilograms of milk per capita (1993) in the developing countries is about three to four times less meat and five to six times less milk than that consumed by people in the OECD countries. Even in 2020, the per capita consumption in the developing world would amount to only 29 kilograms of meat and 63 kilograms of milk per year (Delgado and others 1999), or one-third the level of consumption in the industrial world.

urbanites increases the incidence of chronic diseases. Consumer education and reduction of producer and consumer subsidies is therefore recommended.

Environment

A greater livestock population and increased milk and meat production, coupled with the increased degree of processing that more affluent consumers require, heightens the pressure on global natural resources.[7] While the effects are widespread, the LEAD Initiative (de Haan, Steinfeld, and Blackburn 1997) mentioned several hot spots that require particular attention.

Land Degradation in the Arid, Semiarid, and Subhumid Areas

Traditionally, livestock and overgrazing have been closely associated with land degradation and desertification in the arid areas. However, over the past 15 years or so, a debate has arisen that questions the fundamental principles of rangeland dynamics and pastoral development efforts (Behnke, Scoones, and Kerven 1993; Leach and Mearns 1996; Niamir-Fuller 1999; Sandford 1983; Scoones 1994).

The fundamental principle of the so-called nonequilibrium ecosystems in the arid areas is that climatic variability is of such importance that livestock producers need to be able to "trek" available forage or browse for their animals, which usually requires that they have access to large areas that encompass a diverse range of landscape niches. This calls for livestock mobility and flexibility in access to resources to the maximum extent possible. Subdivision of rangelands and assigning property rights to individuals or groups at too small a scale risks creating rigidities that preclude opportunistic, feed-tracking strategies on the part of resource users. Moreover, convincing evidence is now available that arid ecosystems are extremely resilient (see, for example, Behnke, Scoones, and Kerven 1993; Tucker and Nicholson 1998).

Indeed, precipitation seems to explain a higher amount of variation in vegetation dynamics than previous stocking rates. Flexibility, sustaining mobility, and development of better institutions to manage drought and other forms of covariant risk include some of the key characteristics for the next generation of pastoral projects, such as the World Bank–funded Kenya Arid Lands Resource Management Project, the

---

7. For more specific data on the effect of livestock production on the environment, see de Haan, Steinfeld, and Blackburn (1997).

Morocco Rain-Fed Area Development Project, and the Mongolia Sustainable Livelihood Project.

The situation is much more critical in the semiarid and subhumid regions. Increasing ruminant populations and crop and urban encroachment in the higher potential grazing areas in the semiarid Sahel, West Asia and North Africa, and the Southern Cone of the Americas represent a problem. Along with new investors in livestock, these factors increase the threat of land degradation by overgrazing in those areas (ICARDA 1999; Sapelli 1993). Many other factors lead to negative effects from livestock on the semiarid and subhumid environments—macroeconomic policies, such as overvalued exchange rates, subsidized inputs, crop subsidies, and land-use policies that seek to settle migratory populations, individualize land and water access rights, and maintain artificially high livestock populations through subsidized concentrate feeds for drought mitigation. The West African Pilot Program seeks to introduce a holistic resource management approach in agro-pastoral communities in seven Sahelian countries. The midterm evaluation of the program showed strong enthusiasm by the pastoralists, who reported an improvement in the vegetation, livestock yields, and their quality of life from the project. Quantitative data still must confirm these beneficiary perceptions.

Livestock in Imploding Farming Systems

Ever-increasing population pressure in high-potential areas, especially in the tropical highlands, leads to livestock being crowded out, which, in turn, leads to a downward spiral of reduced soil nutrient transfer and, hence, reduced income opportunities. The strong relationship between the long-term decline in the livestock to human relationship and the deterioration of the diet, income, and ethnic relations in central Rwanda (de Haan, Steinfeld, and Blackburn 1997) is one example. Consequently, policies, institutions, and technologies that encourage the transfer of nutrients, either through grazing animals or animal feed, and discourage the cultivation of marginal areas, that is, the abolition of crop import protection or crop production subsidies, should be promoted.

Livestock and Biodiversity Competition and Integration

Expanding demand for food, including meat and milk, makes expanding, and even maintaining, protected areas for wildlife and other forms of biodiversity increasingly difficult. Livestock and wildlife, therefore, increasingly use the same areas but are not necessarily in competition. Traditional institutional arrangements, involving different public sector agencies and the almost total exclusion of the local population from the benefits of biodiversity conservation, adversely affect the possible synergies that can be generated between livestock and biodiversity conservation. Dietary overlap and the risk of disease, such as rinderpest (Grootenhuis, Njunguna, and Kat 1991), can be managed through preventive vaccinations and judicious management of species composition. Greater participation by local populations in the planning, costs, and benefits, and closer cooperation among the agencies involved, are key requirements to achieve a situation in which livestock and wildlife will thrive together.

Deforestation in the Humid Tropics

Since the 1960s, about 200 million hectares of tropical forest have been lost, mainly through conversion to cropland and ranches, the latter especially in South and Central America. This has caused considerable loss of biodiversity and global carbon emission. Some of the earlier reasons for deforestation, such as subsidized credit, export promotion for beef, and tax incentives for ranch development, have disappeared in almost all Latin American countries. Legal procedures for property rights often still provide the wrong incentives (titles are not issued for forested land) and might encourage deforestation and land speculation, but the overall role of land speculation in deforestation of Latin America's rainforests might be overstated (Faminov and Vosti 1998).

More important, regional demand for food now seems to be a major driving force in land clearing. Many of the rainforest areas are poorly integrated in the national road networks and must rely on their own food production to feed the important current indigenous populations. Deforestation is now more induced by slash and burn practices for food production (maize and soy, plantain), and livestock arrive only after soil fertility is depleted. Key focus areas to reduce deforestation would therefore include (a) a review of the subsidy and legal frameworks; (b) an increased focus on the methods of intensified land use, thereby reducing the pressure for more land to satisfy local food needs; and (c) identifying incentive schemes to promote ecological services such as carbon sequestration and biodiversity.

Waste Production and Intensive Forms of Production

The strong growth of industrial systems of livestock production clearly increases pressure on the environment. Managing mitigation of that pressure is probably the greatest challenge the profession faces. This pressure manifests itself in several ways.

Waste production. Nutrient surpluses from production using feed concentrates, seen earlier mostly in the eastern United States and northwestern Europe, are now also common in areas of East Asia and Latin America. For example, Ke (1998) shows extremely high (more than 800 kilograms per hectare) nitrogen surpluses around the urban areas of eastern China. A rough estimate indicates that about 100,000 square kilometers in the developing world are already threatened by severe nutrient loading, causing eutrophication of waterways and subsequent damage to aquatic ecosystems.

Gas emission. Animal waste produces nitrous oxide—one of the most aggressive greenhouse gases—and ammonia, which in turn cause acid rain and the destruction of marginal landscapes and habitats.

Significant demand for feed grains increases the need for cultivation. More cultivation causes additional erosion, loss of plant and animal biodiversity, and puts an additional strain on the world's scarce water resources. Delgado and others (1999) estimate that under the normal demand scenario, the additional feed grain requirements are about 240 million tons, which, at an average yield of 6 tons per hectare, would require 40 million hectares of additional arable land.

Requirement for genetically uniform stock. The industrial system and the consumer require uniformity, which contributes to an erosion of domestic animal diversity as local breeds are crowded out by industrially popular breeds. The consequent narrowing of the genetic base also increases vulnerability to epidemics.

Many current technologies could mitigate those negative effects. The main issue is a policy framework to induce those technologies. Some of the main policies recommended in LEAD are the following:

Internalize environmental costs in the price of the product. Although more information needs to be collected on the environmental costs of industrial production units, some figures from Australia and Singapore point to a 10–15 percent direct surcharge to mitigate water and soil pollution and abate gaseous emissions (de Haan, Steinfeld, and Blackburn 1997). The key issue will be governments' willingness to impose these surcharges on predominantly urban consumers.

Search for the tools (zoning, taxation, and so on) that will provide a better geographic distribution of intensive production. The key challenge of intensive production is to bring waste production in line with the absorptive capacity of the surrounding land. In particular, pig manure has a high water content, and neither drying nor transporting it over large distances is economically attractive. A combination of zoning regulations and fiscal incentives might be necessary, as is now successfully being tested in East Asia under LEAD.

Promote the use of technologies that increase the efficiency of feed conversion, reducing inputs and nutrient emissions. A large number of technologies currently exist that could improve the digestibility of key nutrients, thereby reducing the emission of nitrogen and phosphates. The adoption of such technologies should be encouraged.

Beyond the environmental effects specific to production systems, many global effects, (the global commons) such as carbon dioxide emissions, increased cultivation of arable land, and erosion of domestic animal biodiversity, can also be detected. Finally, the industrial system might also produce a positive effect—it could reduce the pressure on more fragile ecosystems with unique biodiversity and reduce greenhouse emissions because intensive production systems produce less carbon dioxide per kilogram product than low-production systems (de Haan, Steinfeld, and Blackburn 1997).

The Poor

The trend of intensification and industrialization can potentially have both positive and negative effects for the poor. For the rural poor the issue is mainly the sustainability and profitability of their farming enterprise, but for both rural and urban poor, the cost of basic staples is also affected.

Delgado and others (1999) argue that the livestock revolution might be good for the poor. This projection is based on the assumption that it will be easier for the poor to improve their income when the overall pie is growing in a sector where they already have a major stake, and intensification might provide economically viable alternatives for their labor other than just crops. However, the increased intensification of livestock production, especially the increased concentration and vertical integration of input supply, production, and processing that accompanies it, could provide a serious threat to smallholder production, since they may not be able to compete with the more efficient industrial units.[8]

Considerable economies of scale exist, especially in poultry and pig production, housing, input procurement, and disease control. However, the costs of antibiotic resistance and its spillover into public

8. Commercial producers argue, however, that smallholders, who primarily produce for home consumption but also market their surplus, provide unfair competition.

health, the cost of unsustainable feeding practices (as evidenced by the recent mad cow disease problem in Europe), and occasional large epidemics such as the recent outbreaks of classical swine fever and foot-and-mouth disease also demonstrate the drawbacks of large, intensive production units. Experience from the Bank-funded Philippine credit programs of the 1980s, which financed smallholder broiler production (500–3,000 birds), showed that they were crowded out by multinational enterprises. Business failures among small-scale farmers have been proportionally greater than among large-scale farmers (World Bank 1983). By contrast, the experience with the Turkish Development Foundation (box 2.2) shows that cooperative management of a vertically integrated chain can provide the required framework for economic growth of smallholder producers.

Livestock and Consumers

The role and voice of consumers have become more important in recent decades. This role not only involves food safety (see below), but also animal welfare, sustainability, and public acceptance of production systems. These factors enhance public interest and require greater responsibility of producers and investors.

Globalization, increased consumer concerns, and the emergence of diseases as a result of intensified livestock production lead to an increased

Box 2.2. Vertical Integration and Smallholders: The Example of the Turkish Development Foundation

The Turkish Development Foundation (TKV), largely funded by private sources, has been developing and managing various programs and projects in rural development, and played an important role in the start-up of medium-size poultry producers. TKV's poultry program during the late 1980s and early 1990s helped farmers to

- Establish small poultry farms (capacity up to 4,000 broilers per cycle),
- Set up local and regional cooperatives, and
- Obtain inputs, such as feed, day-old chicks, and medication.

The TKV programs are based on local initiatives, establishing sustainable local organizations with a focus on community, organization, women, livestock improvement such as milk collection, bee-keeping, and so forth. They made a substantial contribution to the development of Turkey's poultry production.

focus on food safety standards in industrial exporting countries. These standards must be an immediate requirement if a primary objective is access to worldwide markets. For developing countries, the stakes are important—food exports can contribute to rural growth and poverty alleviation, but they are extremely costly. Local consumers in the Bank's client countries are the prime losers in the above-mentioned changes in the food chain.

Traditionally, most consumers are aware of food safety risks and take appropriate action to purchase and consume safe food. Good cooking and a direct personal relationship and short distances between producer and consumer prevent food-borne diseases. However, with increasing urbanization and changing consumer behavior, these tools become less reliable. Food poisoning and parasitic and other types of diseases are gaining importance and visibility. More than 2 million children die each year in the developing world from food-borne diseases. Moreover, poor food hygiene contributes to higher health care costs, reduced workforce productivity, and diminished quality of life. Outbreaks of food-borne diseases may also damage a country's tourist industry. Food-borne illness is likely to present a significant problem in developing countries in the future (Kaferstein and Abdussalam 1998).

Good sanitary practices, standards, and compliance with health and food safety obligations are also critical for export development. Food safety is emerging as the most prominent source of conflict in international markets, and poor countries need to upgrade their capacity if they want to maintain access to international markets.

Increased intensification is leading to the emergence of new diseases. Bovine spongiform encephalopathy, caused by recycling animal waste, is a direct result of the increasing scarcity of feed resources and the cost of waste treatment. The reoccurrence of classical swine fever and foot-and-mouth disease, which has led to massive destruction of animals, is directly related to animal densities that increase the chance of infection. Another example is the Nipah virus, which caused a new form of viral encephalitis in Malaysia and led to the destruction of more than 1 million pigs.

Animal Welfare

Unbridled development of industrial production systems—high-density batteries for broilers and layers and sow tethering for intensive pig production—is likely to induce the use of livestock rearing techniques unfriendly to animals. These practices will be phased out in the European Union over the next decade, and they will become an increasingly important issue in the political economy of international development support and international trade. Moreover, the increased climatic variability and recurrent drought induces great animal suffering. Finally, the breakdown of social cohesion in many societies often causes an increased mistreatment of animals. By contrast, concern about animal welfare is not yet high on the agenda of many development planners. Rather than directly imposing Western standards, a more efficient approach seeks policy changes that promote the internalization of negative environmental externalities and thereby encourages animal-friendly forms of smallholder farming. In addition, promoting farming and certification of products that are produced humanely and sustainably might also provide an incentive for better farmers and herders.

Implications for the World Bank

These trends demand a new analysis of the role of the public sector and the means by which it delivers goods and services and, by extension, the role of international funding agencies such as the World Bank. Involvement of the World Bank needs to be seen within its mission to alleviate poverty. And for the World Bank's Rural Sector, its involvement must be seen as part of the mission to alleviate rural poverty within the framework of sustainable natural resource management.

Future interventions in the subsector should emphasize these objectives and be viewed within the institutional context presented above. To summarize, smaller public sector budgets, along with a better understanding of the comparative advantages of the public, private for-profit, and not-for-profit sectors, could lead to smaller (although better defined and strengthened) roles for public institutions. Higher demand for animal foods and resulting intensification could increase the need for public sector involvement in protecting the poor, mitigating negative environmental effects, protecting the consumer, and ensuring food security.

From past emphasis on public sector intervention to increase meat and milk production, the focus will therefore need to shift in order to

- Ensure that livestock maintains its critical role in poverty alleviation;
- Mitigate the negative and enhance the positive effects of livestock on the environment, and promote technologies friendly to animal welfare;
- Develop appropriate institutions and infrastructure to improve cost-effective food safety; and
- Ensure household, national, and global food security, covering food availability, accessibility, and nutritional adequacy.

3

Main Interventions

Ensuring an Appropriate Policy Framework

Many of the policy measures affecting the livestock sector are macroeconomic. They are outside the direct influence of decisionmaking by the technical ministries of our client countries and the rural development specialists of donor agencies. Wider economic policies might be introduced, which may either support or be in conflict with the social and environmental objectives of the sector. A good example is the devaluation of the Communauté Financière Africaine franc (CFAF) in West Africa, which did more for destocking the overgrazed, subhumid areas in Burkina Faso than many of the pastoral development programs (Vergriette and Rolland 1994). However, macroeconomic policies are considered outside the scope of this book, and this chapter will deal only with policies related specifically to the livestock sector.

Financial Policies

Ideally, prices for commercial services and commodities should reflect direct and indirect costs in order to reflect the correct market signals. This approach encourages efficient resource use and, as a general rule, will benefit the environment and the poor.

In the past, the livestock sector has been rather protected in most parts of the world. In the formal plan economies, it was the most heavily supported sector because of its political significance (the level of meat consumption was the measure of well-being in those countries). In Western Europe, Japan, and the United States, red meat and milk are highly protected. In many countries meat and milk prices are still sensitive politically, and in countries such as India and Indonesia where most food prices are now free, governments continue to strictly control milk and meat prices. These policies generally have a perverse effect on the environment. The special import tariff on cassava for intensive pig production in the Netherlands offers an example (Steinfeld, de Haan, and Blackburn 1997).[1] A shift to an open-market-based pricing system is therefore recommended and should be the basis for policy work in the livestock sector. Some important results of such a shift would include the following:

Changes in the international competitiveness of developing countries. More work is required on the subject of international competitiveness, but a more open market might cause a certain redirection of global feed and livestock flows to the developing countries, in line with the projections of Delgado and others (1999). McCalla and de Haan (1998) project that high production costs in Western Europe will produce less, offering opportunities to the areas with lower production costs.

For milk, which can best be produced in regions with high-quality pasture, a temperate climate, and significant capital investments, the obvious first candidates for increasing production will be New Zealand and Australia—the world's most efficient producers. However, all grazing areas in New Zealand are already used, and water constraints will seriously affect the expansion possibilities for intensive dairying in Australia. With long pasture seasons in the temperate zones of the Americas—the United States, eastern Canada, and the Southern Cone of South America—these areas could be the main sources of future growth in milk. There should also be expansion possibilities in some central European countries if and when they can muster the necessary capital and skills for the required processing capacity. This growth will be supplemented by pockets of opportunity for temperate highlands, such as in East Africa, and in areas where dairy production is completely integrated into the cultural fabric of the society, such as India and Pakistan.

For red meat, the more extensive grazing areas of the Americas—the United States, Latin America, and Oceania—will supply most of the deficits, but increased niche opportunities might arise for regions such as the Sahel, which will be in a better competitive position for the West African coastal markets. Similarly, East Africa might improve its competitiveness in Middle Eastern markets. For the savannas of South America and Africa, excellent opportunities present themselves if they can control some current sanitary problems.

---

1. Special import tariffs on cereal substitutes, combined with high protection of domestic cereals in the European Union, caused massive imports of cassava chips, and resulted in soil fertility depletion in Southeast Asia (Thailand) and nutrient loading and environmental degradation in the Netherlands.

Intensive pork and poultry production is less land dependent and can be expected to locate near the consumer. A relocation of intensive production to the developing world can therefore be expected.

In defining regional strategies, more attention needs to focus on these areas of comparative advantage in addition to the risks of (a) intensive overproduction in certain locations and (b) risks and animal welfare aspects presented by the movement of animals over longer distances.

Internalizing environmental costs. Surprisingly little information is available on the effects of internalizing environmental costs. Steinfeld, de Haan, and Blackburn (1997) report a range of 6 to 9 percent incremental costs on the basis of experience in Malaysia and Singapore (pigs) and Australia (beef). These costs do not include the indirect costs of increased cultivation. These cost increases might shift the balance somewhat to less intensive production, but more analysis is required on the potential for such levels to significantly reduce industrial production in favor of mixed farming and other environmentally benign forms.

Introducing cost recovery for livestock services. This is a key issue in the international debate on the provision of livestock services to the poor. It affects animal health and breeding services, and grazing and water fees.

Most animal health (clinical interventions, noncompulsory vaccinations, sale of veterinary pharmaceuticals) and all animal breeding services (selection and multiplication of improved breeding stock, semen production, and artificial insemination) are private goods, and thus they can be efficiently delivered by private providers (Umali, Feder, and de Haan 1992). The benefits from these services can be exclusively captured by the livestock farmer; other farmers cannot simultaneously benefit from the service.

Although research on the distributional consequences of the privatization of livestock services (both veterinary or breeding services) is limited, extensive experience is available on at least partial cost recovery of livestock health and breeding services. Leonard's study in Kenya (1984) found that farmers are willing to pay for reliable and effective services. Total availability of services increased significantly, and the poor gained greater access to the services when cost recovery was introduced. A study by Georges and Nair (1990) in Kerala, India, found that there was no systematic variation in the use of breeding services according to size of landholding. Of farmers owning less than 0.1 acre, 50 percent of the crossbred cows were inseminated using artificial insemination compared to 45 percent of the crossbred cows owned by farmers with greater than 2.5 acres of land.

De Haan and Bekure (1991) showed that in the Central African Republic the smaller herders used more veterinary inputs per head of livestock than the large owners. The introduction of charges for services related to increased demand for veterinary services has also been observed in Cameroon, Chad, and Mali (Ackah Angniman 1997). Lastly, in 1995 a survey of 135 producers in Kenya found a 93 percent satisfaction rate for services rendered by private veterinarians. Furthermore, 73 percent of these producers considered the prices charged to be acceptable (Wamukoya, Gathuma, and Mutiga 1997). The most extensive work has been carried out in three states of India, where it was found that the poor are quite willing to pay for breeding and health services, although somewhat less than the wealthier part of the population (see box on p. 50; Ahuja and others 2001). However, other studies are less positive. For example, Heffernan and Sidahmed (1998) found that for destitute pastoralists in particular private community-based animal health programs may provide little benefit to a long-term secure livelihood. With the drive for cost recovery, it appears that many poor pastoralists are purchasing inadequate amounts of drugs at inflated prices. Thus for the communities involved, accessibility may be improved, but affordability is not.

Grazing and water fees have frequently been proposed for communal areas, even on an incremental basis, with owners of larger herds paying a higher fee per livestock head (Narjisse 1996). The payment of operation and maintenance fees is now common practice in providing pastoral water. Full cost recovery, including construction costs, would have a beneficial effect on grazing pressure and should be pursued. Grazing fees, however, have faced implementation difficulties and have been impeded by political forces; they need careful design with an emphasis on long-term tenure, valuation, and major collection and management responsibilities at the local level with local reinvestment to maintain the resource.

Institutional Policies

Two types of organizations have been promoted: (a) pastoral, which enables organization of a normally marginal and widespread population and (b) processing and marketing organizations, which are of special relevance because of the perishable nature of the product. The experience with these organizations is as follows.

Pastoral. Pastoral organizations represent a principal feature of pastoral development in Africa and West Asia. During implementation they have been rather successful in mobilizing producers around their preferred inputs (animal health, water development) as shown by projects in Senegal, the Central African Republic, Mauritania, and Guinea. They have been less successful in mobilizing pastoralists in resource management, and most associations have not proved sustainable after project completion (Pratt, Le Gall, and de Haan 1998). In conjunction with continued adjustment of project objectives as feedback from project beneficiaries becomes available, more attention to participatory approaches is necessary. Adjustment of the form and size of the associations to the objectives—different organizations for animal health than for water management, for instance—is also necessary.

The situation in Central Asia presents a special case in that a clear decline in social cohesion is evident at the higher levels of organization, with a strong suspicion of cooperatives, especially those with public sector involvement. A fragmentation and return to the family or small group of families as the decisive unit is apparent in many instances. Many of the customary pastoral institutions have shown themselves to be remarkably resilient after decades of state intervention; however, even they have not emerged unscathed from successive collectivization and economic liberalization. This fragmentation has weakened the input of the pastoral sector in consultative processes.

Production, processing, and marketing. These organizations have been successful in Bank-funded projects dealing with the Indian Dairy Cooperatives (box 3.1) and the Turkish Development Foundation (box 2.2). Both groups have been successful in mobilizing large groups of small producers and providing an essential outlet for the produce of the rural poor. The Indian Dairy Cooperative model has been mentioned as one of the most successful commodity programs helping to alleviate poverty, with multiple beneficial effects, including nutrition, education (especially for girls), and job creation (Candler and Kumar 1998; LID 1998). In several states, however, the Indian Dairy Cooperative has not been able to limit excessive government intervention. Both organizations benefited from strong leadership.

Producer organizations can be a crucial factor in advancing the public agenda, although, perhaps with the exception of India and Turkey, they have not been particularly effective in the policy dialogue with the government. Empowerment of rural livestock keepers has not been too prominent in the Bank's operations because in many societies livestock keepers are marginalized in the policy dialogue. With the great increase of civil society participation in drafting policy documents such as the PRSP, this is one of the most important challenges confronting the sector.

More recently, grassroots producer organizations have played a role in priority setting on research, extension, and credit to make government more efficient and responsive to clients. For example, Senegal's recently approved innovative, Agricultural Services and Support to Producer Organizations, has strengthened producer organizations (including herders),

Box 3.1. Operation Flood: How a Commodity Project Can Reduce Poverty

Operation Flood was supported from the mid-1970s to the mid-1990s by the Bank and other donors. It originally started as a marketing project but gradually developed into production and input services. It is based on three-tier cooperative systems that include

- Village-level dairy cooperative societies, which are farmer controlled, with an elected management committee, including at least one woman;
- Regional milk producers' unions that own the dairy plants and transport equipment for milk collection and processing; and
- State federations for interstate sales and coordination.

The National Dairy Development Board, a government organization, provided the technical support. Operation Flood now counts about 9 million members (60 percent are landless), with a daily milk throughput of about 30 million liters. It has made important contributions to poverty reduction, human health, and nutrition. It is the most successful Bank operation in the livestock sector, and arguably in the rural sector. Important issues are interference by the government, in particular in the federations, and its search for monopoly positions now that its support from outside sources is phased out.

providing funds to bring to bear the necessary leverage on service providers (inputs, extension, and research) through a demand-driven services fund that producer organizations can access on a matching grant basis for their training, advisory, and research and development needs.

Ensuring Access to Pastoral Land

For several decades the World Bank has supported pastoral and rangeland development in arid and semiarid areas, particularly in Africa and the Middle East. The manner in which land tenure and policy have been approached in this field, however, has often been found wanting (Bruce and Mearns 2001). Conventional approaches to property rights, balancing the number of grazing livestock against long-term carrying capacity, and assigning property rights to individuals or the state have exacerbated the conflict over resources. These factors have contributed to the acquisition of land of higher potential by elites, uncontrolled privatization of common pastures, and severe environmental degradation.

Although exceptions exist (box 3.2), in many national and Bank-funded activities dealing with land policy, legislation still tends to be drafted with a focus on sedentary crops and an unwarranted need for national harmonization of the legislation. This occurs despite increasing evidence that land titling that results in individual parcels and appropriation of land assets customarily held under corporate (common or communal) ownership may have an adverse outcome on productivity, equity, and sustainability (Baland and Platteau 1998).

It is important that in future livestock and national resource management projects in the dry areas, this issue be confronted explicitly, and clearer distinctions should be made for situations in which individual or group approaches to land titling may or may not be appropriate (Bruce and Mearns 2001). Where land titling (whether individual or group-based, or both) is thought to be inappropriate, alternative policy options need to be developed as ways to promote productive, equitable, and sustainable rural land use. Where relevant, these options should be integrated into the CAS. Three of the key immediate requirements in this area include the following:

a. Develop good practice notes on appropriate land and water access rules, conflict management based on ecology and resource attributes, character of prevailing property regimes (open-access

Box 3.2. Mongolia and Publicly Owned Pasture

In Mongolia a full 80 percent of the total land is publicly owned pasture, held as de facto commons by herding communities. This is perhaps the largest contiguous area of commons in the world. The productive, equitable, and sustainable management of common land is not of marginal concern in Mongolia—it is of central importance to economic growth and poverty reduction in this transition economy.

Mongolia is unique among countries with significant pastoral populations in having a legal framework that actively supports nomadic pastoralism. The 1992 constitution explicitly prohibits the privatization of pastureland, and the 1995 land law upholds customary forms of pastureland tenure, such as seasonal rotation of pastures. This legal framework has evolved in a political-economic context in which herders constitute most of the rural population, and their share of total population has actually been growing following breakup of agricultural collectives in the early 1990s.

Source: Hanstad and Duncan (2001).

public land, controlled-access public land, common property, private property, or a combination of these, for instance), nature of land markets (degree of segmentation), and the presence of patterns of discrimination (social exclusion) toward particular social groups.

b. Create much greater awareness by national policymakers and staff of the Bank and other international organizations about the specific needs for land legislation in dry land areas.
c. Develop new approaches to the management of climatic variability and risk in pastoral areas, such as support for emergency early off-take, management of critical feed and stock resources (grazing reserves, cow-calf camps), closer integration of arid and higher potential areas with feed and grazing lease markets, livestock insurance, and postdrought restocking.

Ensuring Access to Knowledge

The development and dissemination of knowledge to improve animal production and processing has been one-sided. Disease control has been the centerpiece of livestock research and development. This was fully justified when epizootic diseases were the main constraint.

Now the sector faces new challenges: (a) with finite land resources, the increasing demand for animal food products requires further improvement in productivity per animal and per hectare; (b) intensification, which implies an increasing demand by producers for access to knowledge; (c) lack of sustainability in production systems, which implies the need for better waste and natural resource management; (d) reduction of grazing areas, which requires a shift in rangeland management; (e) increased globalization and, therefore, disease risks, which requires more attention to epidemiology; and (f) reduction in budget resources. This book argues, therefore, that in research and extension the public sector should focus increasingly on those areas in which public intervention is required: poverty alleviation, environmental effects of livestock development, and food safety issues.

Research

Livestock research has received little attention in most countries. It has often performed poorly compared to other research areas such as crops and soils. The current situation in national livestock research systems in most client countries of the Bank is one of erroneous priorities, with the focus mostly on "modern" production systems and on-station research. The impact of international research has also been slow to arrive.

Past achievements. There have been several bright spots. On the disease side, achievements include the recent development and application of reliable and user-friendly vaccines in South America against foot-and-mouth disease, in Asia against Newcastle disease in poultry, and in Sub-Saharan Africa against rinderpest. Drugs, such as Ivermectin®, were making a significant difference in parasite control, but drug resistance is an increasing problem. Removing the threat of these devastating diseases opens up new avenues in production, reduces losses, and because of their user-friendly nature, can be further catalysts in promoting producer organizations. In addition, simple application technologies such as tsetse traps and pyrethroid-based vector control have also made impact (Le Gall, de Haan, and Schillhorn van Veen 1995).

On the animal feeding side, an improved understanding of nutrient requirements and the technology to compose balanced rations have made a great impact on feed conversion and nutrient emissions in areas where excess nutrients are a problem. Enrichment of straw and other low-quality feeds with urea has reportedly been adopted on a large scale in China and to a lesser extent in other countries. Similarly, the development of high-yielding, high-quality fodder has the potential to greatly affect efficiency and alleviate poverty in subtropical environments (CIPAV 1998).

On the genetic side, clearly the introduction of artificial insemination and a better understanding of breed ¥ environment interactions have made an enormous difference. Improved genetic selection methods have also contributed to a 50 percent improvement in feed conversion in pigs and poultry in countries such as Brazil and Thailand.

Focus on traditional and indigenous knowledge has improved the application of new technologies and the dialogue between herders and service providers (McCorkle, Mathias, and Schillhorn van Veen 1996).

Future focus. A clearer distinction between the public and private sector roles is also required for research and can be expected to continue evolving. Privately sponsored research will concentrate on those items for which the benefits can be captured by individual investors, such as products that can be patented or their intellectual property rights otherwise protected. This would specifically cover veterinary pharmaceuticals, feed additives, and other products, including biotechnology. The future focus of the public sector on "public goods" will require an emphasis on poverty alleviation and environmental management.

For poverty alleviation the emphasis would include the following:

- Agro-ecologies in which poor people live (marginal areas), which implies more research using animals in nutrient cycling (agroforestry, rotation, manure management, and so on), pastoral land management systems (box 3.2), and high-yielding fodder species for marginal lands.

- Species kept by the poor (small ruminants and smallholder pigs and poultry); use and genetic improvement of indigenous, more resistant breeds; and diseases that particularly affect these species in smallholder situations, such as *peste de petits ruminants*, African swine fever, rabbit hemorrhagic disease, and Newcastle disease in poultry.
- Products and processes adapted to the delivery and marketing systems of the poor, such as thermostable vaccines, low-cost vector controls, appropriate artificial insemination techniques (such as room-temperature semen), long shelf-time products such as dried meat, and cottage processing of milk.
- Integrated farming systems to develop technologies for low-input farming systems, with maximum nutrient recycling, including improving low-quality fodder and energy recycling.

Many of those techniques are also beneficial or benign from an environmental management point of view, and the search for those win-win situations is, of course, highly desirable. Of particular importance on the environmental side (de Haan, Steinfeld, and Blackburn 1997) are grazing, mixed farming, and industrial systems, and global overlays.

For grazing systems, more research and demonstration are needed on the ecological services of pastures (integration with wildlife, carbon sequestration, and so forth), risk diversification in pastoral systems (drought management, income diversification), and intensification of pasture production in humid regions. Major breakthroughs would be achieved with improved digestibility of low-quality fodders and environmentally benign control methods for vector-borne diseases. In mixed farming systems, the emphasis must be on nutrient and energy flows through increased biomass production, reduction of nutrient losses, and increased production efficiency.

For industrial systems, the focus must be on reducing excretion of nitrogen and phosphates by using improved feeding systems, more balanced feed composition, improved diet digestibility, and manure storage and management. Looking at global overlays, such as greenhouse gas emissions and biodiversity, more research would be required on conservation of local breeds to counteract the pressure exerted through intensification, feed improvement, and manure management to reduce the emission of greenhouse gases.

Biotechnology will also play an important role in meeting the increased demand for meat and milk (box 3.3). It offers opportunities for

Box 3.3. Biotechnology and Livestock

Several reproductive and genetic technologies are of interest to the developing world in order to improve the rate of genetic gain.

- *Artificial insemination* is an accepted technology, with about 5 million inseminations in the developing world per year. It still has constraints dealing with the sustainability of delivery to the poor, as discussed in this paper.
- *Embryo transfer* is still a high-cost technology. Multiple Ovulation Embryo Transfer, which focuses on rapid multiplication of outstanding stock, can be relevant for smallholders in developing countries if the selection focuses on the combination of increasing productivity and rusticity, but it is risky if the multipurpose characteristics of the animals are crowded out.
- *Genetic conservation* through cryo-preservation of embryos and semen is now a well-established technology that also has direct relevance to developing countries.
- *Cloning* has only recently shown to be practicable on a wide scale. It potentially threatens loss of genetic diversity and disease resistance, but it can provide significant benefits in developing countries in multiplying, for example, the F1 cross breeds in which, with current techniques, initial hybrid vigor is lost in consecutive crosses.
- Preparation of *genome maps*, which can identify single genes with traits of economic importance or quantitative trait loci that contribute to the variation in economic traits, are increasingly more advanced.
- Accompanied by *marker assisted selection* using genome linkage maps to practical selection methods, this set of technologies can be of critical importance in selecting, for example, for disease resistance in trypanosomosis and possibly in African swine fever.
- *Genetic identification* is relevant for parenthood verification and, more important, in developing countries for determining the genetic distance between breeds, enabling the establishment of priorities for the conservation of endangered breeds.

A recombinant hormone product of interest is bovine somatotropin, a natural hormone in livestock that can be produced relatively cheaply in bacterial cultures and has been shown to increase milk production and disease incidence; this product has now been outlawed in most countries because of too many uncertainties.

(box continues on following page)

Box 3.3 continued

Feed enhancement through rumen manipulation shows promise, although results have been disappointing. As Cunningham (1999) points out, animal feed and the rumen flora and fauna have evolved in parallel, and it is not easy to disrupt or intervene in this system.

In animal health, two areas are of particular interest:

- *Diagnostics,* especially rapid pen or cow-side diagnostic tests for early disease detection, immune levels, and so forth. Such tests are increasingly used in the developing world.
- *Robust vaccines,* with the current search for tropical environments focusing on vaccine development for vector-borne blood parasites, and thermo-stable vaccines that can be used with a minimum of infrastructure by nonprofessional staff.

Sources: Based on Cunningham (1999) for reproductive and genetic technologies and ILRI (1999) for animal diseases.

more efficient production and healthier products, but also poses threats to the environment, social equity, animal disease resistance, and consumer health. These threats necessitate the development of a more articulate strategy on priorities, safeguards, and so on. Such a strategy would need to cover scientific areas of support and consumer concerns about biotechnology products.

For scientific areas of support, public funding would need to focus on the development of pro-poor technologies, which because of limited markets and poor producers, would not be of immediate interest to the commercial sector. Examples are work on the genetics of disease resistance, especially the use of Marker Assisted Selection, exploring disease resistance to African Swine Fever, vaccine development for vector-borne and zoonotic diseases, and biological disease control. Technologies to improve the conservation of local breeds and the improved use of tropical forages would fall in the same categories. Areas where public support would not be justified include the work on drug development and approval (including bovine somatotropin), artificial insemination, and embryo transfer because they generally benefit the wealthier parts of the population.

The key areas of consumer concern about biotechnology products are the ethical aspects of cloning and the use of bovine somatotropin and the increased use of antibiotics it might entail. Compared to plant transgenesis, the introduction of genes from another organism into livestock is more costly, deals with more complex traits (compared to the single-gene disease resistance in plants), and because of the much longer generation interval, is commercially less attractive. However, recent advances in gene mapping that will assist the use of transgenic animals for the production of modified milk (for example, production of pharmaceuticals) might spark a strong ethical debate. These areas of concern need to be included in the national debates on priority setting.

Institutional framework. An institutional strategy to strengthen national livestock research systems would include preparation of national livestock research plans that clearly reflect the economic and social constraints of the sector with clear targeting, priorities, and mechanisms to express demand and coordinate research; emphasis on institutional pluralism (universities, nongovernmental organizations [NGOs], private sector, on-farm research, and so on) and promotion of private-public interaction and competition; encouragement of regional interaction among researchers ("south to south"); improved research quality by linking researchers in the developing world to relevant research and methodologies in industrial countries and systems, including links to international research systems; decentralization research activities to production areas; increased producer involvement in research governance and funding (see producer organizations in institutional policies); and support for competitive funding.

Extension. In most countries livestock extension is provided by numerous players, but it has been neglected in the official extension system. In traditional government livestock services, animal health control was the main focus, but because many agents in the agricultural extension services were biased toward crops, that area received the most attention, and livestock extension became the "filler" of the system.

Over the years the World Bank has promoted the training and visit system, which concentrates on the public sector and argues for an integrated crop and livestock service with a single extension agent at the village level. The system was to introduce stricter organizational rules in the management of staff with a single line of command, time-bound tasks, and a single focus on extension (without other activities such as input supply, data collection, or distribution of loans).

For the dissemination of livestock technology, the results have been poor. The single line of command often conflicts with the separate ministry or department of the extension agent. The single focus on extension often conflicts with the additional animal health tasks expected by the department (and farmers) from the livestock agent, and with the flexibility needed in livestock farming. The animal health tasks are also more attractive financially for the agent. In addition, the time-bound nature does not fit with livestock information needs, and more flexibility is needed (Morton and Matthewman 1996). It is therefore not surprising that livestock extension scored low on the farmer evaluations (although high in their needs) carried out by the extension system in Burkina Faso (Bindlish, Evenson, and Gbetibouo 1993). Finally, the training and visit system is expensive and not financially sustainable.

Combining current experience and the overall focus of this book, the model(s) to promote livestock extension would have the following characteristics.

Multiple sectors: A wide range of sectors (public, private, civil society) should be involved in providing technical information—governments, universities, NGOs, cooperatives, other producers, paraveterinarians, traditional healers, private veterinarians (box 3.4), and commercial enterprises delivering inputs and purchasing and processing products. For the pure pastoral grazing systems, development needs to be based on pastoral organizations. Moreover, for more intensive systems the skills and

Box 3.4. Contracting Livestock Extension to Private Veterinarians: The Case in Mali

Private animal health networks provide a good opportunity to improve livestock production extension. Private veterinarians like to diversify their activities and increase clientele and income, and governments need to support the private sector and reduce the costs of their extension services.

Projet d'Appui au Secteur Privé de l'Elevage, funded by the French Cooperation, supports livestock extension in Mali through a network of private veterinarians and paraveterinarians, with strong participation by producer organizations and regional agriculture chambers (RACs) in programming and contracting. The RACs have contracted 120 private agents to provide training in 4,500 villages (two visits per year). The private agents are paid by the number of villages they visit. The RAC is in charge of paying the private agents, controlling on the basis of a survey carried out in 15 percent of the villages. One false record is enough to cancel the whole due payment and contract. The costs are only US$20 per village per year. Extension messages were provided free to farmers. Vaccines, drugs, and other material such as fly traps had been charged at full cost by the private veterinarians. In 1999, for the most promising region with 40 private agents, results include the following:

(box continues on following page)

Box 3.4 continued

- 1,850 villages received advice on poultry diseases and 2,500,000 birds were vaccinated against Newcastle disease.
- 1,250 villages received information on sheep and goat diseases and 15,000 small ruminants were vaccinated.
- 175 villages attended training on trypanosomosis control and 500 fly traps were sold; and
- 425 villages were trained in range and water management.

Factors affecting the success are

- Strong participation by the herders in selecting priorities;
- Development of extension programs with rapid and tangible effects such as a decrease of mortality rates;
- Extension messages associated with improving access to commercial inputs (drugs, vaccines, traps, feed, and material); and
- Individual contracts for each private agent to ensure appropriate logistics, refresher training, and strict control.

Sustainability is still an issue, with the continuing willingness of the village communities to pay as the key factor.

knowledge of the "unique" extension agent are inadequate, and specialized systems require enhancement. Vertically integrated systems along the commodity chain, although with the potential to crowd out the small farmer, can be an effective change agent in the sector.

Multiple sources of finance: Livestock extension for the special focus areas of natural resource management, poverty alleviation, and food safety is a public good and should therefore be funded through the public sector, as is currently the case. The private operators mentioned in the previous paragraph, if involved in these areas, should be subcontracted through the government. However, for the private-good services of increasing production, cost-recovery must be introduced, and can partially occur through the sale of inputs or the purchase of production, although conflicts of interest might occur. Individual extension services targeted to wealthy producers should in all cases be on a cost-recovery basis.

Multiple channels and media: Traditional extension services have emphasized field demonstrations, oral communication, and some use of radio. For pastoral production, the classic demonstrations of some improvement for individual animals did not make sense. In addition, information requirements for pastoralists are different and often lie in the areas of resource access (grazing, water), movement restrictions, and the means of dealing with public administration.[2] The information technology revolution has provided great opportunity to broaden the mode in message transmission. Communication can still best take place through other media (radio and television) and other places (markets and so forth), but Internet access in rural areas is rapidly increasing, especially in the former Soviet Union, the Middle East, and Latin America. For more intensive production systems, shared learning between farmers and farmer-to-farmer visits have also been extremely effective.

Decentralized decisionmaking and farmer empowerment: With the increasing decentralization of government in many of the Bank's client countries, extension services need to be decentralized and the responsibility for organization brought to the client level. Farmer empowerment should be a critical aspect of the livestock extension strategy, as it should be for any public agricultural services. Farmer empowerment should not be restricted to the local level, but it should be extended to regional and national levels. Some pastoral organizations such as Federation Nationale des Eleveurs Centraficain in the Central African Republic and organizations in Guinea, Mauritania, and Mali are starting to become effective interlocutors with the government. However, farmer and herder perspectives have limits, and an obvious need exists to counterbalance the mostly short-term perspectives of farmer groups with longer term sustainability needs. For instance, water development in arid regions is one of the most necessary and requested services in the ongoing World Bank–funded Arid Lands Program in Kenya, although past experience clearly shows that water development leads to increased settlement and land degradation.

Targeted technology and information: The increased focus on the poor, natural resource management, and food safety issues should be reflected in the technology being extended. This implies a heightened emphasis on the following:

- Small ruminants and smallholder pigs and poultry, focusing on the dissemination of (a) disease control technologies such as the use of heat-tolerant Newcastle vaccines and *peste de petits ruminants* control, with a focus on integrated control programs; (b) improvement of local breeds and use of simple breeding methods; (c)

2. Roe, Huntsinger, and Labnow (1998) consider pastoralists a "high reliability institution," well adapted to attain peak performance and capable of managing highly complex technology to better manage risk. Such institutions are (a) in search of information to increase reliability, especially promoting pastoralist-to-pastoralist links; (b) in need of links to other high reliability institutions (abattoirs, veterinary services) to manage peak load production; and (c) in need of access to scale-dependent resources.

development of fodder packages that give high yields per hectare of high-quality feed; and (d) on-farm processing. Others might include the technologies mentioned in the sections of this book dealing with research.

- Land improvement in marginal areas through the promotion of technologies such as agroforestry and waste management in nutrient surplus areas.
- Assistance in developing "advocacy messages" for marginal groups and enhancement of farmer skills in dealing with resource access, and administrative and regulatory issues.
- Nonlivestock and even nonfarm employment opportunities. Clearly, for many groups such as pastoralists, a strong need exists for out-migration and employment generation outside the sector.

Search for more innovative incentives for extension agents: Extension agents generally had to respond to central state agencies, not to local governments, agencies, or farmer organizations. Salary and other benefits for extension agents have traditionally been determined in the rather rigid framework of the public sector. However, for a strongly decentralized operation such as an extension service, a more performance-based and quantitative incentive system would be preferable. Systems such as share farming with farmers on demonstration plots, payment according to productivity (piglets born per sow per year), and payment according to the quantity of vouchers collected from farmer-clients have been proposed but only barely implemented in Bank or other projects (Ameur 1994). Increased efforts would be required.

Client orientation for extension agents: Extension agents often derive from urban or educated rural societies and frequently consider themselves highly superior to the level of their peasant clients. Requesting that the agents survey indigenous knowledge often leads to improved mutual respect and improved interaction.

Education. An increasing awareness is emerging that technology transfer cannot be accelerated through extension or farmer organizations without focusing on basic and advanced education. Prolonged effects of successful skills development programs begun in the 1970s (mainly through bilateral programs, especially within the U.S. Agency for International Development) have hidden this gap, but it is now evident that the knowledge gap between industrial and developing countries is widening and that skills training programs in most developing countries are inadequate. Training for nomadic people is a special case (box 3.5). In the livestock sector this is exemplified by the significant decline over the last 20 years in the skills in animal health and nutrition, breeding,

Box 3.5. Nomadic Education: A Special Challenge for a Pro-People Livestock Strategy

As part of the preparation for this book, a special review of nomadic education was prepared, which suggests the following:

- Most education services are ill-adapted and often even undermine the pastoral system and the potential for endogenous change.
- Successful education depends more on a context sympathetic to nomadic culture than on the strategy, methodology, or curriculum.
- Nonformal approaches, delivered in a nonantagonistic fashion, matched by pastoral development strategies that help to decrease labor intensity and free up children have been more successful. This means that nonformal approaches should
  - Value pastoral livelihood systems as appropriate and adapted to the environment;
  - Be based in part on indigenous or local expert knowledge and be linked to wider features of social organization; and
  - Recognize that the children have to be equipped for life in other environments, but this might not be the main objective of their schooling.

Source: Kratli (2001).

and so forth, despite a great increase in the number of university "trained" graduates. The net result is an oversupply (especially in the Europe and Central Asia and Middle East and North Africa regions and Latin America and the Caribbean) of poorly trained and poorly functioning veterinarians.

The lack of sufficient skills is especially relevant in exporting countries where animal health staff are responsible for food safety and certification of export products. Inadequate skills in quality control may have serious repercussions for public health in addition to a country's market share in international markets.

Privatization of veterinary services may in the long run help to focus animal health skills, but this needs to be backed up by an overhaul of skills development in animal health and animal production, especially at the university level. This would require projection of skill needs, such as various models that exist for the number of veterinarians required in different types of animal health delivery systems; development and acceptance of a set of basic requirements (this is in place for veterinarians but often overruled by political consideration or lack of resources to fund adequate training programs); consolidation of training programs and facilities to an absolute optimum, reflected by adequate training resources, training skills, access to training materials, and necessary infrastructure; acceptance of international standards in training, skills, and laboratory testing; and introduction of new learning tools, including distance learning, use of simulation models, and so forth.

Ensuring Access to Financial Services

Pro-poor financial services are essential to enable the rural poor to enter into livestock keeping.[3]

Credit

The Bank has wide experience in the provision of credit for livestock. Much of the funding for livestock development in Latin America and Central Europe in the 1970s and 1980s was directed to semipublic ranch and dairy development. The initial aim was to improve genetic stocks. Such projects often emphasized large-scale cattle distribution to "modern" production facilities such as dairy farms or beef ranches. These livestock delivery schemes required sophisticated inputs and veterinary care and were often supported by public sector funding. Moreover, they appeared to concentrate on herd expansion rather than increasing efficiency. In some countries they had a significant impact on the transformation of the livestock sector. In Nigeria in the early 1990s, for example, about 23,000 loans were made totaling approximately US$22 million, boosting the country's fattening industry (World Bank 1996). In other cases the livestock delivery schemes were not sustainable and failed, especially when managed through the public sector, as in Sub-Saharan Africa. Moreover, credit schemes often benefited the wealthy rather than the poor. Directed credit has largely disappeared from the Bank portfolio. Livestock borrowing, however, has not diminished. In Europe and the Central Asia region, for example, lending for livestock often absorbs 40 percent or more of the financial intermediary's portfolio.

Support for smallholder development only came to the development agency agenda in the late 1970s and 1980s with new emphasis on the recognition of women's roles, the importance of traction, and multipurpose use of a variety of livestock species in an integrated smallholder farming system. Such credit may provide essential inputs that significantly

3. This section's main input comes from a Bank-funded review by Jennifer Schumacker of Heifer Project International (Afifi-Affat 1998) and from Schillhorn van Veen (2001).

increase productivity (improved genetics, milking equipment, shearing equipment, feed supplements, medicines, animal traction equipment) and marketability of the final products (milk coolers, wool balers, small-scale processing equipment), or reduce waste (feeding troughs, composting tools). A good example is the Asian Development Bank-funded Bangladesh Livestock Development Program (box 3.6).

In general, however, livestock credit programs require similar treatment as other poverty-oriented credit programs and should therefore be dealt with outside this particular livestock development strategy.

In-Kind Provision of Credit

In-kind livestock schemes are particular to the livestock sector. Traditionally, they have been a part of wealth transfer in most parts of the world, either as (pre-) inheritance, assistance after calamities such as drought or epidemic disease, or informal risk avoidance or insurance schemes. In many parts of the developing world, this remains the case, although many of these traditional systems are breaking down. Most donors financed in-kind livestock schemes in development projects built on these traditions and on the associated societal oversight over such schemes (box 3.7).

Box 3.6. Asian Development Bank Bangladesh Participatory Livestock Project

One of the project's components is microcredit for livestock enterprises. The government of Bangladesh will lend US$17.2 million equivalent to the Palli Karma-Sahayak Foundation (PKSF)—the apex microcredit organization under the project—at a service charge of 1.25 percent per year for a period of 20 years, including a grace period of 5 years. PKSF will lend to participating NGOs at no more than 6.25 percent, with a repayment period of up to three years. The spread of up to 5 percent will cover evaluating and selecting NGOs, monitoring their performance, training NGO staff in financial management, credit risks, profit, and other operating costs.

The participating NGOs pass the proceeds of the loan from PKSF to smallholder poultry, beef-fattening, goat-rearing, and other small animal operations at the prevailing market rate, currently 16–20 percent, keeping a margin of at least 9.75–13.75 percent to cover social mobilization, group formation, and social awareness training, skills training, monitoring loan collection and extension activities, and allowance for bad debts, profit, and other operating costs. Under PKSF's current guidelines, service charge rates to participating NGOs are in the range of 2–5 percent, and rates charged by NGOs to the final borrowers are based on market rates.

Box 3.7. In-Kind Credit on Java

The Provincial Development Program of Central Java Province, Indonesia, introduced a new in-kind loan project in the mid-1980s in order to replace the existing small ruminant credit system. The target farmers were divided into groups of 10 with each farmer receiving two female goats or sheep. The leader of each group received training in small ruminant management and also received a good quality buck or ram.

Each recipient had to repay four lambs or kids of 8 months old over a period of 3 years. Post-program evaluation in 1988 demonstrated that this project could be used to introduce new technology, increase farmer income, improve the production performance of existing goats and sheep, and improve group dynamics of farmer groups.

The overall objective of the in-kind credit system is to provide poor farmers with access to necessary goods and means of production when formal or informal credit schemes fail. The latter is often associated with the lack of access to livestock or new farmers or farmers under new conditions (for example, transmigration farmers in Indonesia, or new farmers in the former Soviet Union). The concept of in-kind livestock credit systems is based on the provision of live animals to poor farmers who then have an obligation, within a given time, to return a set number of offspring.

Factors that contribute to the effectiveness of implementation. There are four factors (Afifi-Affat, 1998):

a. The skills and interest of the implementation agency in livestock development and poverty alleviation are vital. Most common agencies include NGOs and (public sector) livestock departments. Financial institutions have not been particularly effective carrying out in-kind livestock schemes unless the institution has proven its ability to supervise village development.
b. Trust and good relations supersede all technical aspects in ascertaining success. The groups most likely to succeed were women's groups that had been in existence for at least three years. The most challenging were groups whose main purpose was to apply for immediate assistance.
c. The implementing agency's attitude toward decisionmaking and delegation of authority is important. Steady guidance and willingness to give authority to communities appears to be a guiding principle for the implementing agency, as are transparent details

on various arrangements for eligibility, pay-back requirements, and penalties for defaulters.

d. Management information systems and good communication, especially at the community level, are essential. Experience indicates that the most successful communities have good and strictly enforced bylaws and a plan for leadership rotation (generally two years).

Eligibility and selection of recipients. One of the major problems of most targeted credit schemes, whether cash or in-kind, is that it does not or only partially reaches the target population. Various examples exist where more than one-half of credit earmarked for rural areas found its way into urban-based communities that did not really need it. The advantage of in-kind livestock credit is its relatively low fungibility. The risk of reaching wrong targets can be reduced further by careful selection of recipient families (rather than single owners). This is mainly achieved through social control such as participation in groups before and during the in-kind provision.

Ex post performance analyses. Most ex post analyses have been favorable. The Indonesia Smallholder Cattle Development program, cosponsored by the Bank and the International Fund for Agricultural Development offers a good example. The program showed an ex post economic rate of return of 16 percent (Afifi-Affat 1998). More than 70,000 families received about 80,000 cattle of the Bali or Madura breed. The overall development objectives of establishing nucleus herds in transmigration sites and improving the agricultural income of transmigrants were largely achieved. The in-kind credit disbursement and repayment system worked well. Farmer repayments were consistent with average calving rates, although full repayment periods (average seven years) have been longer than anticipated at appraisal (five years), which was too optimistic. At project completion, 65 percent of farmers due for repayment (the project ended before all beneficiaries completed their in-kind payment) had made full repayment (two calves). This credit repayment performance is better than most other credit operations in the agricultural sector (World Bank 1994).

Ensuring Access to Animal Health Services

Adequate access to animal health services is critical for poor livestock keepers, who are often geographically dispersed, in remote areas, and because of their small stock numbers, do not get attention from government services.[4]

Background

In many countries, structural adjustments, reduced government budgets, stronger separation of public and private sector roles, and increased emergence of diseases related to intensification have already changed the way animal health services are financed and provided. In Africa the quality of animal health services, which were wholly under the responsibility of the state immediately after independence, began to deteriorate in the late 1970s, with serious economic consequences. The example usually cited is that of rinderpest. The cost of this disease reappearing in Sub-Saharan Africa in the early 1980s was estimated to have been US$500 million (FAO, quoted in Gauthier, Siméon, and de Haan 1999). Animal health services have shown, almost globally, a rapid growth in staffing, which has not been accompanied by an increase in operating costs so that the staff can work efficiently.

There was a need for more direct interventions as the focus changed from mass control of diseases to individual treatments, and livestock ownership almost worldwide shifted from herders and farmers with experience to new entrepreneurs or smallholders with less knowledge. A reform campaign was thus initiated. Since then, considerable progress has been made in the redefinition of the role of the public and private sectors. Table 3.1 presents a summary of the different roles, as described by Umali, Feder, and de Haan (1992). In interpreting the table, a clear differentiation should be made between responsibility and implementation. Many of the tasks described in the table can be carried out by private operators under the supervision of the public sector.

The outcome is different in various countries and regions, but in some African countries where services were privatized, the channels for livestock services at the micro level have become more diverse and include professional veterinarians, village veterinary pharmacies, auxiliaries, veterinary public and private paraprofessionals, private formal and informal drug and vaccine sellers, and traditional healers. This seemingly open market is, however, not yet perfect. A low density of producers, low effective demand, the often short list of products used and available, and the administrative delimitation of the

4. Mostly based on Schillhorn van Veen and de Haan (1995) and Gauthier, Siméon, and de Haan (1999).

Table 3.1. Classification of Animal Health Services by Public or Private Sector Activity

Service	*Private sector*	*Public sector*	*Economic characteristics of service*
Clinical services	Yes	No	Mainly private good
Production of vaccines and other veterinary products	Yes	Rarely	Mainly private good; some epizootic vaccines have public good values
Distribution of veterinary products	Yes	No	Mainly private good
Vaccinations and vector control	Yes	Possibly	Mainly private good, but there may be externalities[a]
Elimination of animals with epizootic disease	Yes	Possibly	Removal of disease spreaders may have public good aspects
Diagnosis	Yes	Possibly	Mainly private good, but there may be externalities
Veterinary research	Yes	Yes	Private good or public good
Education and extension	Possibly	Yes	Mostly public good[a]
Policy development	No	Yes	Public good
Disease surveillance	No	Yes	Public good[a]
Quarantine	No	Yes	Externalities[a]
Quality control of veterinary products	No	Yes	Information asymmetries
Food hygiene and inspection	No	Yes	Information asymmetries[a]

a. These public good services can be contracted to the private sector.
Source: Adapted from Umali, Feder, and de Haan 1992.

veterinary systems may lead to effective local monopolies (or oligopolies involving paraprofessionals and auxiliaries). These factors may also create a principal or agent relationship, where the market is not always able to distinguish between high and low quality. Better information dissemination, professional self-regulation (veterinary "chambers"), and some public sector oversight may help improve quality and market transparency.

Decentralization has added an additional feature to the definition of a public good. In most countries with a decentralization policy, the control of contagious diseases has been entrusted to the lower (district) level. This is appropriate if a common interest in control exists throughout the entire area, as shown by the decentralized Bank-funded disease control project in Brazil. Working with farmer associations, foot-and-mouth disease was successfully eradicated in two states. Market failures arise if an unequal level of interest prevails. For example, the control of a disease may be seen as a public good at the national level or by most of the country (for the country to be free from a highly contagious disease, for instance). If addressed at a local level, however, it may not represent a priority, perhaps because of a small livestock population.

On the contrary, the control of a disease may no longer represent a public good at the national level, such as an enzootic disease with a low economic effect nationally and no impact on human health, but it could remain a priority at the local level. The responsibility for control must then be delegated locally.

Past performance. Representative quantitative information on the impact of these measures and the performance of the new private operators is still lacking. Some examples, partly from a survey (Gauthier, Siméon, and de Haan 1999) by the directors of veterinary services in Africa, revealed that most countries reported a decline or at least stabilization in the number of civil servants and a reduction of the salary burden on the overall budget; the policy of cost recovery is clearly underwritten by most decisionmakers, even though certain vaccinations and laboratory diagnoses are still largely subsidized; the number of private veterinarians rose in Sub-Saharan Africa from almost none in the mid-1980s to 1,965 in 1995, and most decisionmakers consider that privatization has had a positive effect on the ability of livestock producers to access veterinary care and products; community animal health workers are increasingly recognized as an integral part of the veterinary delivery system, and, although sustainability and quality issues are relevant, most experiences are positive; and private veterinarians are increasingly being used to subcontract public services such as epidemiological surveys, compulsory vaccinations, and food inspection.

Based on his experiences in Senegal, Ly (2000) warns that the newly privatized animal health market could evolve into a typically inferior market, where only low-quality services would be demanded.

Indeed, restructuring of public services is felt by many decisionmakers, particularly in Africa, which has caused a decline of the livestock sector in the overall rural development agenda. Transferring part of veterinary activities to the private sector or to a joint advisory service has reduced the amount of information available because of a lack of communication between the various participants. Conflicting evidence on whether the restructuring has benefited the poor (Gros 1994; Leonard 2000). Ahuja and others (2001) shows that the poor are willing to pay for health and artificial insemination services, although less than the wealthier farmers (box 3.8). Finally, the emergence of a private paraveterinary group is believed to lead to dangerous situations of inappropriate drug use and to eventually threaten human and animal health, which indicates that some audit and oversight are required.

Long-term needs. Quite clearly, the long-term objective of any effort to improve the delivery of animal health services is to provide quality services at the lowest cost. This argues, especially in areas with a lower livestock density, for a combination of the professional veterinarian and a livestock farmer, technician, or grassroots auxiliary. This is the model in many industrial countries and could also be increasingly supported in the developing world. The sale of veterinary pharmaceuticals directly to livestock owners or auxiliaries, along with advice on their use, becomes an increasing part of the income of a private veterinarian.

Box 3.8. Livestock Services and the Poor: The Case in India

An in-depth survey in three states in India on the effect of commercialization of veterinary and artificial insemination services showed the following:

- Government and private veterinarians charge similar fees for health and breeding services. The formal low fee of the government services was charged to only about 20 percent of the cases, of which between 10 and 40, depending on the state, benefit the poorest quintile of the population.
- The poor are willing to pay but only to a limited extent. For example, according to a contingent valuation method, the poorest quintile is willing to pay up to Rs 600 (about US$10), whereas, the wealthiest quintile is willing to pay up to Rs 1,000–1,200, depending on the state.
- The elasticity of demand for curative services and artificial insemination is very low. The services are valued very highly, and price is not foremost in making a decision.

Source: Ahuja and others 2001.

This approach would require the following actions:

a. Continued support for privatization of services is necessary by creating an enabling environment for the private animal health service provider. Full cost-recovery for private good services provided by the public services would be introduced, along with increased subcontracting of public sector tasks to private veterinarians.
b. Increased emphasis would be on training farmer-veterinary auxiliaries, which would do such tasks on a part-time basis for hire. The auxiliaries could receive their training from private veterinarians who would continue to provide them with technical advice and inputs.
c. An adjustment in the legal and regulatory framework for delivery of veterinary services would be necessary. This would reflect changes from a past situation with scarce and, therefore, government-qualified veterinarians supported by para-veterinary staff, to the current desirable model of private veterinary practice with community animal health workers or veterinary auxiliaries. Experience in Benin, Côte d'Ivoire, Guinea, and Madagascar indicates that establishing modern legislation that openly recognizes the private practice of veterinary medicine contributes significantly to developing and improving veterinary services. A legal framework is necessary to avoid rent seeking, especially in areas with a strong public sector. The Office International des Epizooties has drafted model veterinary pharmaceutical legislation that only needs to be adapted to local circumstances.
d. Continued strengthening and restructuring of public services is essential—performance-based salary structures and increased training in public responsibilities (epidemiological surveillance, quality control of veterinary drugs, food safety control methods, application of standards for international trade, health economics, and so forth).
e. Greater attention to risk analysis and cost-benefit considerations in adapting animal health policies.
f. Use of low-cost control methods such as integrated, internal parasite control screens; traps for tsetse control and heat-tolerant vaccines would be needed (Schillhorn van Veen and de Haan 1995; Le Gall, de Haan, and Schillhorn van Veen 1995).
g. Increased support for transparency of animal health markets and creation of an oversight mechanism, either by the public sector or through self-regulation by the private sector, would also be necessary.

Thus the future strategy should continue its two-pronged approach—support for public-sector functions, including newly emerging tasks such as food safety, and continued privatization and subcontracting of public sector tasks. Private animal health care should focus on a network of professional veterinarians, paraveterinarians, and trained livestock farmers. This could provide an efficient and low-cost system of animal health care.

Ensuring Access to Breeding Services

In many developing countries animal breeding is in the hands of government institutions that increasingly employ artificial insemination and on-station cross-breeding with exotic breeds. While promoted by many donors (including the Bank), government institutions have frequently been ineffective and sometimes counterproductive in improving the genetic level of smallholder livestock. Two of the key issues include the following:

Exotic Genetics

Breeding strategies that involve introduction of exotic genetics are often not adapted to local environmental conditions. The promotion of modern breeds, often ill-adapted to the local environment but sustained by feed and other subsidies, still prevails. Such genetics normally benefit the wealthier part of the population. Incorrect performance monitoring (for example, using annual yields rather than lifetime yields and using biological yields rather than efficiency or production costs) is a part of the cause. Moreover, genetic diversity and long-term breeding goals might be threatened by such policies. The situation is serious, with some of the best tropical breeds in danger of extinction. The FAO's Global Databank for Animal Genetic Resources indicates that, globally, 19 percent (in the developing world, 21 percent) of the approximately 4,000 livestock breeds are in danger of extinction. Particular hot spots include Chinese pig, South Asia dairy, and some of the noncattle bovid species. Where well-structured programs are in place, however, such as in dairy development in India and East Africa, the introduction of exotic breeds makes a significant contribution to rural welfare.

Artificial Insemination

Artificial insemination, requiring sophisticated equipment and flexible management, is ill-suited for situations with poor infrastructure and rigid rules of public administration. The developing world has about 5 million artificial inseminations per year, excluding the former Soviet Union, mostly in Asia (Cunningham 1999). Although artificial insemination, in addition to embryo transfer, is a mature technology, conception rates of artificial insemination in developing countries are less than 50 percent almost without exception, thus defaulting on the main objective of most poor farmers—obtaining regular offspring. In addition, the cost per insemination is high, and the increased output in many tropical conditions remains low, so that continued subsidization and promotion are needed to keep such programs alive. This keeps out private artificial insemination companies and stifles innovation.

Given the limitations described in the two preceding paragraphs, the following actions are required:

a. An enabling environment that provides a level playing field for different breeds should be established. Too often, the incentive framework favors exotic breeds at the expense of local breeds, with subsidies for concentrate feed, artificial insemination, and veterinary services. A distorted incentive system will negatively affect environmental sustainability and social equity.
b. Conservation of local breeds should receive increased attention. At the international level, many activities relating breed characterization deserve further support. These activities are spearheaded by the FAO and the International Livestock Research Institute, in addition to numerous private breed preservation groups or associations. The World Bank, which mostly works at the national level, directs its support for local breed conservation through genetic improvement and development of the production systems. Several projects (India, Ghana) with village-based selection using the open nucleus breeding system are being supported, but it is too early to assess their effectiveness. First experience indicates that it is difficult to focus the attention of government officials away from the classic government ranch approach to village- or community-based selection schemes.
c. Sustainability and the economics of breed improvement with exotic breeds should receive more attention. This includes attention to the costs and benefits of genetic improvement, including calculating lifetime yields versus annual yields, quality and price of the genetics, and the sustainability of the infrastructure (liquid nitrogen and so on). Subsidies should have a clear sunset clause.
d. Increased attention to other services such as animal health is also necessary.

Ensuring Access to Markets

Access to markets is an important channel for a functioning production complex that is able to manage livestock-environment interactions and food security and safety and better target livestock-related poverty alleviation. Such markets need to be open and transparent to allow producers to rapidly respond to changing demand or market conditions. Most activities are the responsibility of the private sector; indeed, most Bank-funded interventions in livestock marketing through the public sector have failed completely. This section, however, focuses on issues that require public investments.

Physical Infrastructure

Physical infrastructure, such as marketplaces, slaughterhouses, processing plants, and storage facilities, is often needed to improve access to markets. Constructing and maintaining such infrastructure is generally within the activities of the private sector. In the past the public sector has invested heavily in livestock market infrastructure and processing plants to (a) facilitate trade and provide sufficient quantities of processed products and (b) improve hygiene, especially in slaughter and processing.[5] Subsidies for hygienic improvements, including upgrading slaughter facilities, might be justified for countries to access export markets or, as in the case of Eastern Europe, to reach the standards of the European Union, or where significant positive effects on food safety or animal welfare can be detected.

Because of externalities, intervention by the state is justified in this type of market. However, even if the state is involved (whether the central state or lower entities such as counties or municipalities), local associations of traders and butchers must be consulted in the design. For instance, the terminal market in Abidjan was built in the 1960s to a design based on a European model without any involvement of local personnel. Traders refused to use the new facility because of its inconvenient design, and the market has remained unused. Also, investments in markets with relatively low maintenance costs have generally been more satisfactory than constructing huge structures. The Sudan Livestock Marketing Project is a good example. It was a success because it filled a major need in Sudan's livestock transport and marketing systems. The ex post economic rate of return was 40 percent (World Bank 1989).

---

5. Most animal slaughter in 19th- and 20th-century Europe was carried out in municipal abattoirs with a strong emphasis on hygiene and prevention of zoonotic disease. These were slowly privatized or replaced by private processing plants in the late 20th century.

Slaughter facilities range in capacity and complexity from the single outdoor slab to large, technically advanced industrial abattoirs. In most developing countries, slaughterhouses provide a service to the local butchers but do not themselves buy stock and sell meat products. Many countries are facing problems linked to inadequate facilities and overcapacity. For instance, in China capacity utilization varies considerably in slaughterhouses from less than 10 percent to more than 70 percent of the maximum design. Below 70 percent, it is difficult to yield a profit. Similarly, in the former Soviet Union, where the livestock inventory has contracted to approximately 50 percent of its level in the 1980s, there was massive overcapacity of all livestock-related services, including processing.[6]

Other factors causing the lack of capacity utilization include an initial design often inappropriate to the actual market, management by public servants who have few marketing skills, and traditional livestock production patterns that lead to idle periods. A major problem is that the price elasticity of the slaughter services is quite low—in many instances the slaughter fee required to ensure functioning of the facility is so high that many butchers adopt alternative (clandestine) slaughtering. To limit the development of informal processing without quality standards, and in view of the public health and environmental benefits, public sector support of well-managed animal slaughter facilities may be justified.

Privatization has not been frequently successful largely because of a lack of agreement or cooperation between the beneficiaries (that is, the butchers) and low price elasticity of demand for the services. Experience shows that a complete divestiture to the private sector is not feasible, and the search must be for service contracts, management contracts, or leases. One common problem is valuation of facilities. Both China and India control the ownership of land and use land on which their old facilities are located as a share in new joint ownership companies and, thus, a share of any future profits. However, the value of land and existing buildings on publicly owned sites is frequently overvalued by governments, which causes considerable friction between the public sector owners and the private sector project sponsors.

Marketing cooperatives have rarely been sustainable for meat marketing but have been fairly successful in marketing single commodities such as milk, poultry products, honey, silk, or wool. In

6. Most of this capacity is now defunct, exacerbating the problem of poor quality processing in the former Soviet Union and in many parts of Africa. Mitigating this problem would require an almost complete rebuilding of infrastructure. This is increasingly done around large urban centers (by the private sector or by public-private partnerships) but not in rural areas, leaving the rural population with lower quality, higher waste and losses, and greater public health risks.

many cases such marketing cooperatives broaden their mandate and branch out to provide supplies and advisory services.

Transaction Costs and Information Flows

Market development requires efficient exchange relationships and access to information, and many projects have sought to improve the availability of market data. Nevertheless, several generic problems associated with information can be identified: (a) processing information can be a costly activity; (b) information can be manipulated in a strategic manner, leading to opportunistic behavior; and (c) property rights to information may be incomplete, especially in terms of excludability. Because of these market failures and social aspects (market information can easily be monopolized by the wealthy), the role of the public sector is evident. Nevertheless, public sector involvement in this process has generally been disappointing, and better identification of actual information needs is necessary.

Special Market Interventions

Among the major impediments to market access are over-regulation and taxation, and the lack of pragmatic and cost-effective solutions when mingling international standards and local capacity and affordability. In this case the State is more often seen as the problem than as stimulating development. Sound policy advice is needed on an appropriate mix of broadly accepted and practical regulations and encouragement for private initiatives.

Drought management is one of the most powerful arguments for proactive public sector involvement in marketing. Drought has tremendous social and environmental effects, and its occurrence is predicted to increase because of global warming. The main environmental effects occur in arid and semiarid areas during periods of drought in which perennial shrubs and trees that are difficult to restore in the short term die. On the socioeconomic side, losses are high. For example, in northern Kenya drought in the early 1990s led to losses of approximately 40 percent for cattle and goats and 16 percent for camels, estimated at a value of at least US$46.2 million (Barton and Morton 1999). In addition to the direct impact on stock losses, indirect consequences are also important, especially for pastoralists who face strong erosion or total loss of their livelihood. A key characteristic of markets in drought is that the number of stock on the market increases and quality declines (box 3.9). Consequently, the terms of trade erode between animal products and the goods purchased by herders (livestock prices decrease as staple food prices increase).

Box 3.9. Drought Subsidies to Livestock Traders in the Isiolo District, Kenya (1996)

Earlier experience of direct purchase by projects in Kenya indicated that these schemes succeeded only in removing the poorest and weakest animals, for which there was little market, from the range.

During the drought in 1996, the Dutch-funded Isiolo District Project, in cooperation with Action Aid, implemented a new scheme based on a subsidy to livestock traders in order to involve them in the trade. The average subsidy, considering all administrative costs (security for traders who carried cash to sales and administrative costs associated with organizing and arranging markets and providing contagious bovine pleuropneumonia [CBPP] tests) was approximately US$13 for each animal bought under the drought relief scheme. About 3,000 cattle were removed from the range at an average price of US$103. Therefore, the gross benefit to households in Isiolo District was US$310,000. This is equivalent to food relief for 36,000 adult months equivalent (this includes the cost of the food plus administration and transport).

Source: Barton and Morton (1999).

Reviews of drought assistance have almost always come back to the long-term effects of well-intended short-term relief actions. The intervention policies that have shown some effects are (a) encouraging farmers and herders to be more self reliant, (b) maintaining and protecting the environment, and (c) facilitating postdrought recovery.

Intervention in marketing during a drought emergency can maintain the purchasing power of pastoralists and prevent land degradation (Foran and Stafford Smith 1991). An early intervention, for example, through subsidies in animal transport when drought threatens, might be more effective than dealing with the consequences of a drought, which include food aid, social upheaval, and even armed conflict (see box 3.9). This comprehensive assessment, including the cost of no action, must be considered in the cost-benefit ratio.

Sanitary Regulations

Any decision on the level of investment in food safety needs to be preceded by an in-depth analysis. "Modern" food safety measures are expensive and must be evaluated based on their returns to improved public domestic health and export earnings, or both. Too often the desire for modern OECD-type regulations prevails over rational food preparation criteria that consumers demand in domestic and export markets. Risk assessments and cost-benefit analysis should therefore be an essential part of defining a framework for finding cost-effective solutions to food safety problems.

Risk assessment means identifying the sources and importance of different risks and evaluating the costs and benefits for different interventions to reduce them. This is especially true for small countries that may not meet food safety requirements because they lack economies of scale. In addition, these countries might have weak international representation, and support might be needed to become proactive participants in the many international organizations dealing with food safety issues. Their difficulties should be taken into account so that trade liberalization, combined with the growing trend toward stricter sanitary and phytosanitary (SPS) standards, does not exclude them from the benefits, if any, of international trade (Univehr and Hirschhorn 2000).

Food safety is generally managed in the public domain because it involves externalities and moral hazards. However, much of the process and oversight can be contracted to the private sector. Where capacity exists, the best practice is to rely on transparent private certification and investments under the supervision of the state. On the contrary, the introduction of quality grades, particularly if they are linked to brand name development, is a pure private good and should be left to the private sector.

Improved standards for the domestic market must be implemented step by step and follow the pace of consumption changes to keep the system economically sustainable. Thus the role of the state is to provide only basic standards to avoid fatalities. Milk production and its marketing system is a good example to show a potential counterproductive impact related to a high-standard strategy. In many developing countries, preparation practices include cooking and boiling, so health risks related to milk consumption are not as important. Pasteurization of milk would double the costs, depressing the market for smallholders and putting the product out of the reach of the poor. Moreover, the use of a new technique might introduce a false confidence in the safety of sometimes inadequately processed products.

To date, World Bank experience in financing food safety programs is limited. Programs have been mainly in areas of quality control for export in Africa, the Balkans, and Asia (aquaculture). The China Smallholder Cattle Project, which covers the entire food chain from smallholder cattle fattening to clean processing and marketing, including training in Hazard Analysis and Critical Control Point Analysis, offers a good example of the opportunities provided by chain management. In addition, some experience has been gained in setting up food regulatory systems in

Eastern Europe. However, with the growing importance of this area, the Bank may increase its attention and skills to counterbalance organizations such as the World Trade Organization and help countries decide on appropriate and cost-effective food quality control systems that ensure affordable food to local consumers, contribute to rural growth and poverty alleviation, and protect human health.

4

The Bank's Livestock Portfolio: Past, Current, and Future

Since the late 1970s livestock lending supported by the World Bank has been declining (figure 4.1). The volume of lending for livestock-only projects has stabilized at a low number of projects since the mid-1980s. Currently, six active agriculture projects are livestock only, and about 50 projects (of the total agriculture portfolio of more than 270) have livestock components. The decrease in livestock lending was partially related to the poor performance of the projects of the 1970s and 1980s,

Figure 4.1. Average Annual Lending (Total Project Costs) for Livestock-Only and Livestock Component Project Costs Funded by the World Bank since 1974

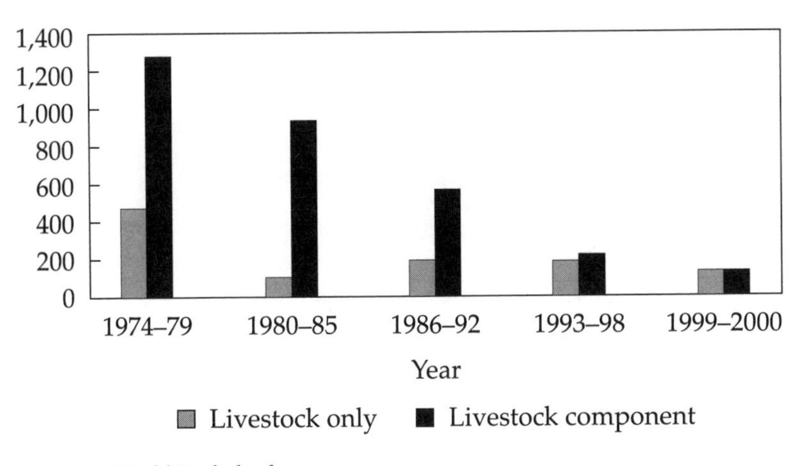

Source: World Bank database.

partly because of privatization of state enterprises such as large production farms and processing facilities,[1] and partly because of a general decline in agricultural lending in the Bank. Indeed, since the mid-1980s direct lending for livestock-related interventions has declined at about the same rate as other agriculture. By contrast, indirect lending through agricultural credit schemes increased.

The performance of the livestock portfolio is equal to the rest of the rural portfolio. Of the livestock-only projects, 75 percent shows a satisfactory or highly satisfactory performance.

Figure 4.2 shows a breakdown of the different livestock-related activities in the active World Bank portfolio of US$1.5 billion. The Bank's livestock lending can be characterized by several factors:

a. The projects are heavily concentrated in the East Asia and the Pacific region, which represents more than one-half of the investments. The regions with a much higher share of rural poor

Figure 4.2. Percentage of the World Bank's Fiscal Year 2001 Livestock Portfolio (US$1.5 billion) as Allocated by Region, Species, and Activity

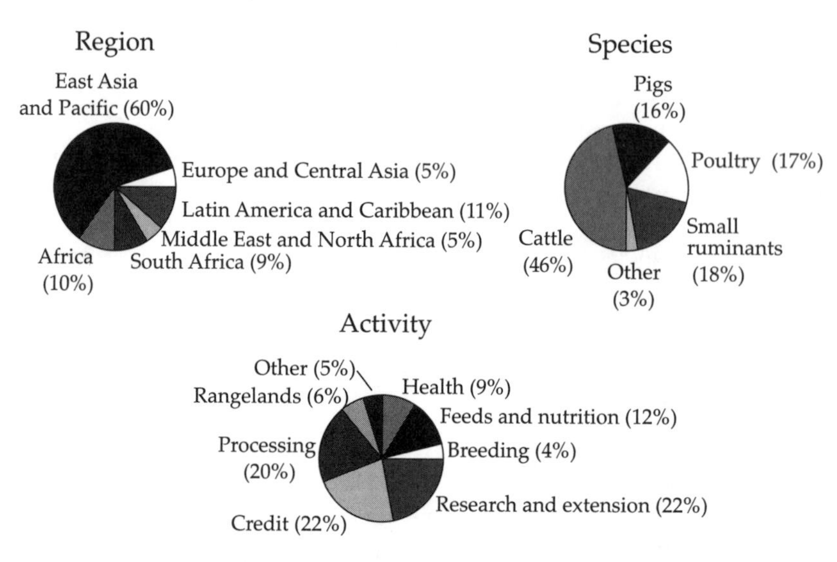

Note: Excludes most onlending for livestock credit.
Source: World Bank data.

1. Financing of livestock production and processing by the International Finance Corporation increased during the mid-1990s.

(Sub-Saharan Africa and South Asia) each have only one-tenth of the World Bank's support for the sector.

b. The portfolio covers the entire food chain, with strong emphasis on credit, collection and processing, and capacity building in information systems. About 20 percent of the current portfolio is directed at market development, supporting capacity building in sanitary and phytosanitary services (Brazil Plant and Animal Health Project, Niger Agro-Pastoral Export Promotion) or improvement in the domestic hygiene and quality standards in meat, such as the China Smallholder Cattle Development Project.
c. The portfolio is still strongly oriented toward improved cattle production, although the share of nonruminants becomes increasingly more important. Nonruminant production is particularly strong in the East Asian portfolio, while Sub-Saharan Africa and South Asia are still heavily oriented toward cattle. Attention to other species (rabbits, sheep, and deer) is minimal and in China and central Asia only.
d. The portfolio is weak in rangeland (and, more generally, in pastoral development) and other environmental issues related to livestock, with few specific investments in the key livestock and environment hot spots such as on-farm nutrient management of intensive production units and livestock and deforestation in the humid tropics.

The Future

The opportunities are exciting: livestock can be a key sector to enable poor farmers to escape the poverty trap; strong demand is growing for livestock products, particularly in our client countries; and the understanding of how to use livestock to reduce poverty has increased. The situation demands policy advice and well-selected investments to assist the poor (producers and consumers) and ensure production sustainability. Great challenges also exist: the livestock sector is under pressure from public opinion and benign neglect from decisionmakers and must demonstrate that the shift recommended toward public goods to reduce poverty, environmental sustainability, and food safety is critical to safeguard future populations. This environment coincides with negotiations among international and bilateral agencies in industrial and developing countries that have widely different perceptions on the priority to be allocated to issues such as the environment and animal welfare. A broad and innovative action plan is required. It would have four components:

COMPONENT 1
Sharpen and Increase World Bank Support to Reduce Poverty and Vulnerability through Livestock Development

For at least two-thirds of the world's rural poor, livestock can be the vehicle through which they can escape the poverty trap. However, the challenges are equally large because of a lack of good practices in pro-poor design of livestock operations and because the nature of public investments is changing. Moreover, a sea change will be required from the international livestock community to make people, rather than animals, the focus of livestock development. Focusing on the multiple functions of livestock and including poor non-livestock keepers as potential beneficiaries of livestock development deserve a higher priority than the former exclusive focus on increasing milk and meat output for urban consumers.

More specific actions include the following:

a. In cooperation with other agencies, develop a deeper understanding of the key aspects of pro-poor design of livestock development operations. An initiative similar to LEAD (box 4.1) may help document and demonstrate viable approaches to improve the livelihoods of the poor through livestock development. This would

Box 4.1. Livestock, Environment, and Development

LEAD was established in 1998 to "protect and enhance natural resources as affected by livestock production while alleviating poverty." It is a 10-donor initiative implemented by the FAO that has provided about US$3.5 million to achieve the following:

- Develop a Virtual Center for Livestock and the Environment that brings together discussions, decision tools, and other relevant information on livestock-environment interactions.
- Train and offer policy advice on livestock interactions through seminars, workshops, and so forth.
- Pilot interventions in the hotspots, that is, livestock and wildlife, the livestock and tropical forest interface, intensive livestock production, and waste management.

The Bank has benefited from LEAD through support in policy dialogue in Madagascar, Morocco, and Turkey; and seed money to develop Global Environment Facility (GEF) project proposals in Central America, Eastern Africa, and possibly East Asia.

also require a closer integration at the Bank level of livestock specialists in environment and in broader sector activities. Similar to LEAD, such a multidonor initiative would provide seed money to start pilot activities and support.

b. Develop training modules and provide training and enhance awareness in pro-poor livestock development strategies for decisionmakers involved in the PRSP and rural strategy formulation.[2]
c. Strengthen Bank support for production systems practiced by most of the rural poor—the small-scale mixed farmers. Applying the lessons of the past in livestock service delivery and integral food chain management, this subcomponent would need to include sector work on the incentive framework for smallholders (eliminating eventual bias for industrial production), increased attention to producer organizations in input supply, processing and closer participation in the policy dialogue, continued attention to privatization of animal health and breeding services, and multiple source advisory sources.
d. Increase support for pastoral development through continued piloting and upscaling along the lines of the "nonequilibrium" principles, paying particular attention to pastoral empowerment, resource access, and drought preparedness, including markets and appropriate financial instruments (insurance, savings) for the mitigation of drought and other risks.
e. Avoid funding large-scale commercial, grain-fed feedlot systems and industrial milk, pork, and poultry production except to improve the public good areas of environment and food safety.

COMPONENT 2
Increase World Bank Support for Management of Livestock-Environment Interactions

The expected increase in the demand for animal products is projected to result in a 50 percent increase in the feed grain requirement, and unless the productivity per animal is greatly increased, similar increases in livestock numbers are projected. This will greatly increase the pressure on the world's natural resources and, without action, continue degradation. Specific actions to be taken include the following:

---

2. For example, the interim-poverty reduction strategy paper draft from Ethiopia, a country with a large rural herder population that received in 1999 more than US$100 million in food aid, devoted only one paragraph to this topic and population.

a. Continue integrating livestock-environment interactions in sector work, national environmental action plans, and so forth. Key issues in the policy dialogue concern incentives for large-scale versus small-scale producers, internalization of environmental costs in price, and common resource access issues.
b. Continue developing and testing innovative approaches to management of livestock-environment interactions in hot spots, such as drought preparedness to address desertification of arid rangelands, introduction of benefit-sharing systems for livestock-wildlife systems, payment for ecological services in degraded pastures to address deforestation in the humid tropics, waste management of processing and production, and areawide integration of industrial units to address nutrient loading and groundwater pollution. In several of these areas, close cooperation with the Global Environment Facility would be a further step in testing and upscaling current experiences.
c. Mainstream sound ecological farming practices, such as integration of crops and livestock, development of markets for organic products, and so on.
d. Reduce waste, including disease and mortality, through a better understanding of disease epidemiology prevention and treatments.

Component 3
Strengthen the Bank's Food Safety Involvement

This new and perhaps growing area for the Bank would include the following actions:

a. Promotion of consumer- and producer-based food safety awareness, based on local tradition and food habits.
b. Regulations including fine-tuning and, where necessary, reinforcing the role of the public sector in regulation and policies.
c. Infrastructure, covering investments in quarantine infrastructure, diagnostic laboratories, contaminants and microbial counts, slaughter facilities, and research infrastructure to improve standards, markets, and sanitation.
d. Human and institutional capacity building for legislation, regulation, and certification, in addition to technical skills and best practices in general food hygiene. Most important, the poor need a strong voice and participation in international standard-setting organizations. Priorities, for example, should ensure that the world market will not inappropriately interfere with internal regulations and controls or unnecessarily increase food costs.

e. Promotion of regional cooperation between small countries to ensure that their interests are included in international standard-setting processes.

COMPONENT 4
Maintain the Bank's Leadership in the International Community

Over the past decades, Bank staff have played a leadership role in the discussion on public and private roles of livestock services, management of livestock and environment interactions, and, more recently, in food safety and pro-poor livestock development design. Bank staff involvement in the LEAD Initiative has led to an increased global awareness of the need for urgent action in identifying policies and technologies that mitigate negative and enhance positive effects. This involvement has contributed to changes in the research and development agenda of institutes such as the European Union, the International Livestock Research Institute, the International Food Policy Research Institute, and the FAO. However, much of the innovative development work is done outside the public-sector-supported international organizations, and these experiences should increasingly be included in implementing livestock development.

Focus on development thinking. The Bank has a comparative advantage because it can combine policy and investments and is seen by many as the leader in development thinking. Bank staff should continue with this role, especially in the short term. Specifically the Bank should continue active support of the LEAD Initiative, particularly in developing the second phase and in mainstreaming results of the first phase into the operations of the GEF and the Bank. It should provide new leadership on understanding the basic characteristics of pro-poor design of livestock development operations and the means of integrating such a design into the policy dialogue (the PRSP and the CAS). Bank staff should pursue new initiatives in the area of diseases, trade, and food safety, particularly in helping countries and donors analyze priorities and formulate country action plans, and provide leadership in mainstreaming results of current initiatives in Bank operations. Finally, it should develop the policy and technology framework to enable developing countries to benefit more from the shift in the industrial countries toward less intensive forms of production.

Focus on strategic staffing. Over the next three years, several Bank livestock specialists will retire. Already, most staff involved in past programs of public sector intervention to increase meat and milk production have retired. Good opportunities exist to develop staffing skills that can respond to the new mandates of focusing international finance and the public sector on the key issues of poverty reduction, skills development, environmental sustainability, and food safety. At the same time, this should be seen in the context of declining rural staff numbers (World Bank forthcoming). Four actions are recommended:

a. Maintain a livestock development adviser at the Rural Development Department with excellent Bank and international credentials who can further develop the international discussion on current critical global issues in livestock development. The adviser would play a key role in promoting livestock development for poverty alleviation by advocating the integration of livestock development in the PRSP, the CAS, and others and maintaining high-quality standards for livestock development in Bank operations.
b. In the regions with strong pro-poor livestock development needs (Sub-Saharan Africa, central Asia, and South Asia), maintain, or appoint, high-level livestock generalists.[3] These generalists, with development and private sector experience, could be the leaders to shape policy dialogue on pro-poor, sustainable livestock development in their respective regions in sector work, projects, and programs. They would be responsible for preparing and implementing such operations, expanding the development dialogue to other sectors (health, private sector development, and so forth), and cooperating with other international organizations such as the FAO and the Cooperative Program.
c. Develop closer cooperation with the International Finance Corporation and, together, show how commercial livestock development can be used for poverty alleviation.
d. Maintain a formal or informal network of livestock specialists.

3. This is especially relevant in those regions (Africa, Europe, and Central Asia) where there has been a national erosion of technical and livestock economic skills over the last few decades.

References

Ackah Angniman, P. 1997. "Privatization of Veterinary Services within the Context of Structural Adjustment in Mali, Cameroon, and Chad." Paper presented at the First Electronic Conference: Principles for Rational Delivery of Public and Private Veterinary Services, Food and Agriculture Organization of the United Nations, January 1–20.

Afifi-Affat, K. A. 1998. "Heifer in Trust: A Model for Sustainable Livestock Development?" *World Animal Review* 91(2): 13–20.

Ahuja V., P. S. Georges, S. Ray, K. E. McConnell, M. P. G. Kurup, V. Ghandi, D. Umali-Deiniger, and C. de Haan. 2001. "Agricultural Services and the Poor: Case of Livestock Health and Breeding Services in India." Indian Institute of Management, Ahmedabad, India.

Ameur, C. 1994. "Agricultural Extension: A Step Beyond the Next Step." World Bank Technical Paper 247. World Bank, Washington, D.C.

Baland, J. M., and J. P. Platteau. 1998. *Halting Degradation of Natural Resources: Is There a Role for Rural Communities?* Oxford, United Kingdom: Clarendon Press and New York: Oxford University Press.

Barton, D., and J. Morton. 1999. "Livestock Marketing and Drought Mitigation in Northern Kenya." Draft report. Natural Resource Institute.

Behnke, R. H., I. Scoones, and C. Kerven, eds. 1993. "Range Ecology at Disequilibrium: New Models of Natural Variability and Pastoral Distribution in Africa." Overseas Development Institute, London.

Bindlish, V., R. Evenson, and M. Gbetibouo. 1993. "Evaluation of T&V-Based Extension in Burkina Faso." World Bank Technical Paper 226. World Bank, Washington, D.C.

Bourne, D., and W. Wint. 1994. "Livestock, Land Use and Agricultural Intensification in Sub-Saharan Africa." Pastoral Development Network Paper 37c. Overseas Development Institute, London.

Brown, L. R., and H. Kane. 1994. "Full House: Reassessing the Earth's Population Carrying Capacity." World Watch Institute, Washington D.C.

Bruce, J., and R. Mearns. 2001. "Land Policy and Natural Resource Management." In K. Deininger and others, eds. *Land Policy and Administration: Lessons Learned and New Challenges for The Bank's Development Agenda*. World Bank, Washington, D.C.

Candler, W., and N. Kumar. 1998. *The Dairy Revolution: The Impact of Dairy Development in India and the World Bank's Contribution*. Washington D.C.: World Bank.

CAST (Council for Agricultural Science and Technology). 1999. "Contribution of Animal Agriculture to Meeting Global Human Food Demand." CAST, Ames, Iowa.

CIPAV (Centro de Investigación de Agricultura del Valle). 1998. "Sistemas Pecurrios Sostenibles para las Montanas Tropicales." Fourth International Seminar, October 4–6,1995. Calli, Columbia.

Cunningham, E. P. 1999. "The Application of Bio-Technologies to Enhance Animal Production in Different Farming Systems." *Livestock Production Science* 58(1): 1–24.

Danish Ministry of Food and Agriculture. 1999. *Action Plan II—Development in Organic Farming*. Copenhagen, Denmark.

de Haan, C., and S. Bekure. 1991. "Animal Health Services in Sub-Saharan Africa: Initial Experiences with Alternative Approaches." World Bank Technical Paper 134. World Bank, Washington, D.C.

de Haan, C., T. W. Schillhorn van Veen, and K. Brooks. 1992. "The Livestock Sector in Eastern Europe: Constraints and Opportunities." World Bank Discussion Paper 173. World Bank, Washington, D.C.

de Haan, C., H. Steinfeld, and H. Blackburn. 1997. Livestock and the Environment: Finding a Balance. European Commission Directorate-General for Development, Brussels.

Delgado, C. L., M. W. Rosegrant, and S. Meyer. 2001. "Livestock to 2020: The Revolution Continues." Paper presented at the International Trade Research Consortium, January 18–19, 2001, Auckland, New Zealand. International Food Policy Research Institute, Washington D.C.

Delgado, C., M. Rosegrant, H. Steinfeld, S. Ehui, and C. Courbois. 1999. "Livestock to 2020: The Next Food Revolution." 2020 Vision Initiative Food,

Agriculture, and the Environment Discussion Paper 28. International Food Policy Research Institute, Washington, D.C.

Fafchamps M., C. Udry, and K. Czukas. 1998. "Drought and Savings in West Africa: Are Livestock a Buffer Stock?" *Journal of Development Economics* 55(2): 273–305.

Faminov, D. M., and S. A. Vosti. 1998. "Livestock–Deforestation Links: Policy Issues in the Western Brazilian Amazon." In A. Nell, ed. *Proceedings of the International Conference on Livestock and the Environment*. International Agricultural Centre, Ede / Wageningen, The Netherlands, June 16–20, 1997.

FAO (Food and Agriculture Organization of the United Nations). 1997. *Statistical Yearbook*. Rome, Italy: FAO.

_____. 2001. *Statistical Yearbook*. Rome, Italy: FAO.

Foran, B. D., and D. M. Stafford Smith. 1991. "Risk, Biology and Drought Management for Cattle Stations In Central Australia." *Journal of Environmental Management* 33(1): 17–33.

Gauthier J., M. Siméon, and C. de Haan. 1999. *The Effect of Structural Adjustment Programs on the Delivery of Veterinary Services in Africa*. Proceeding of the Regional Conference of Office International des Epizooties for Africa, Dakar, Senegal, January 25–29. Paris.

Georges, P. S., and K. N. Nair. 1990. "Livestock Economy of Kerala." Centre for Development Studies, Trivandrum, India.

Goodland, Robert. 1997. "Environmental Sustainability and Agriculture: Diet Matters." *Ecological Economics* 23(3): 189–200.

Grootenhuis, J. G., S. G. Njunguna, and P. W. Kat, eds. 1991. *Wildlife Research for Sustainable Development: Proceedings of an International Conference Held by the Kenya Agricultural Institute*. Kenya Agricultural Institute, Nairobi, Kenya.

Gros, J. G. 1994. "Of Cattle, Veterinarians, and the World Bank: The Political Economy of Veterinary Services Privatization in Cameroon." *Public Administration and Development* 14(2): 37–51.

Hanstad, T., and J. Duncan. 2001. "Land Reform in Mongolia: Observations and Recommendations." RDI Report of Foreign Aid and Development No. 109. Rural Development Institute, Seattle, Washington.

Heffernan, C., and A. E. Sidahmed. 1998. "The Delivery of Veterinary Services to the Rural Poor: A Framework Analysis." Paper presented at the conference Provision of Livestock Services to the Rural Poor, June 12, University of Reading, United Kingdom.

Henry R., and G. Rothwell. 1996. *The World Poultry Industry*. International Finance Corporation Global Agribusiness Series. World Bank, Washington, D.C.

Hu, F. B., and W. C. Willett. 1998. *The Relationship Between Consumption of Animal Products (Beef, Poultry, Eggs, Fish and Dairy Products) and Risk of Chronic Diseases: A Critical Review*. Harvard School of Public Health, Cambridge, Massachusetts.

ICARDA (International Center for Agricultural Research). 1999. *Annual Report*. Aleppo, Syria.

ILRI (International Livestock Research Institute). 1999. *Annual Report 1998*. Nairobi, Kenya.

Kaferstein, F., and M. Abdussalam, 1998. "Food Safety in the Twenty-First Century." Paper presented at the 4th World Congress on Food-Borne Infections and Intoxications, June 8–11. Berlin, Germany.

Ke, Bingsheng. 1998. "Industrial Livestock Production, Concentrate Feed Demand and Natural Resource Requirements in China." In A. Nell, ed. *Proceedings of the International Conference on Livestock and the Environment*. International Agricultural Centre, Ede / Wageningen, The Netherlands, June 16–20, 1997, 180–91.

Kratli, S. 2001. "Education Provision to Nomadic Pastoralists." Working Paper, Institute for Development Studies, University of Sussex, United Kingdom.

Le Gall, F., C. de Haan, and T. W. Schillhorn van Veen. 1995. "Simple Animal Health Techniques." Agriculture Technology Notes No. 10. World Bank, Washington, D.C.

Leach, M., and R. Mearns, eds. 1996. *The Lie of the Land: Challenging Received Wisdom on the African Environment*. Oxford, United Kingdom, and Portsmouth, New Hampshire: James, Currey, and Heinemann.

Leonard, D. K. 1984. "The Supply of Veterinary Services." Development Discussion Paper 191. Harvard Institute for International Development, Cambridge, Massachusetts.

_____. 2000. "The New Institutional Economics and the Restructuring of Animal Health Services in Africa." In D. K. Leonard, ed. *Africa's Changing Markets for Health and Veterinary Services*. London: Macmillan Press.

LID (Livestock in Development). 1998. *Poverty Reduction: A Review of Best Practice in the Livestock Sector*. Crewkern, U.K.: Livestock in Development.

Ly, C. 2000. "Management and the Impact of Auxiliaries on Pastoral Production and Veterinary Services Delivery in Senegal." In D. K. Leonard, ed. *Africa's Changing Markets for Health and Veterinary Services*. London: Macmillan Press.

McCalla, A. 1999. Food Security and the Challenge to Agriculture in the 21st Century. Rural Development Note No. 1. World Bank, Washington, D.C.

McCalla, A., and C. de Haan. 1998. "An International Trade Perspective on Livestock and the Environment." In A. Nell, ed. *Proceedings of the International Conference on Livestock and the Environment*. International Agricultural Centre, Ede / Wageningen, The Netherlands, June 16–20, 1997, 13–21.

McCorkle, C. M., E. Mathias, and T. W. Schillhorn van Veen. 1996. *Ethnoveterinary Research and Development: IT Studies in Indigenous Knowledge and Development*. London: Intermediate Technology Publications.

Mitchell, D. O., and M. D. Ingco. 1993. *The World Food Outlook*. World Bank, Washington, D.C.

Morton, J., and R. Matthewman. 1996. "Improving Livestock Production Through Extension: Information Needs, Institutions and Opportunities." Natural Resource Perspectives No. 12. Overseas Development Institute. London.

Narjisse, H. 1996. "The Range Livestock Industry in Developing Countries: Current Assessment and Prospects." In Neil West, ed., *Proceedings of the Fifth International Rangeland Congress*. Vol. 2. Salt Lake City, Utah, July 23–28, 1995. Denver, Colorado: Society for Range Management.

Niamir-Fuller, M. 1999. *Managing Mobility in African Rangelands*. London: Intermediate Technology Publications.

Pimentel, D. 1997. "Livestock Production: Energy Inputs and the Environment." In paper presented at the 47th Annual Meeting of the Canadian Society for Animal Production. Montreal, Quebec, July.

Pratt, D., F. Le Gall, and C. de Haan. 1998. *Investing in Pastoralism*. World Bank Technical Paper No. 365. World Bank, Washington, D.C.

Roe E., L. Huntsinger, and K. Labnow. 1998. High Reliability Pastoralism. *Journal of Arid Environments* 39(1): 39–54.

Sandford, S. 1983. *Management of Pastoral Development in The Third World*. Chichester, United Kingdom: John Wiley & Sons.

Sapelli, C. 1993. "Inflation, Capital Markets and the Supply of Beef." *Agricultural Economics* 9: 154–62.

Schillhorn van Veen, T. W. 2001. "Livestock-in-Kind Credit." Agriculture Technology Notes No. 27. World Bank, Washington, D.C.

Schillhorn van Veen, T. W., and C. de Haan. 1995. "Trends in the Organization and Financing of Livestock and Animal Health Services." *Preventive Veterinary Medicine* 25(2): 225–40.

Schroeder, T. C., A. P. Berkeley, and K. C. Schroeder. 1995. "Income Growth and International Meat Consumption." *Journal of International Food and Agribusiness Marketing* 7(3): 15–30.

Scoones, I. 1994. *Living with Uncertainty: New Directions in Pastoral Development in Africa*. London: Intermediate Technology Publications.

Simpson, J. R., X. Cheng, and A. Miyazaki. 1994. *China's Livestock and Related Agriculture: Projections to 2025*. Wallingford, United Kingdom: Commonwealth Agricultural Bureau International.

Steinfeld, H., C. de Haan, and H. Blackburn. 1997. Livestock: Environment Interactions: Issues and Options. European Commission Directorate-General for Development. Brussels.

Toledo, J. M., 1990. Tropical Pasture Technology for Marginal Lands of Tropical America. *Outlook on Agriculture* 15(1): 2–9.

Tucker, C. J., and S. E. Nicholson. 1998. "Large-Scale Saharan-Sahelian Vegetation Variations from 1980 to 1996 Derived from Precipitation and NOAA Satellite Data." In V. E. Squires and E. Sidahmed, eds. *Drylands*. Rome: International Fund for Agricultural Development Technical Reports.

Umali, D., G. Feder, and C. de Haan. 1992. "The Balance Between Public and Private Sector: Activities in the Delivery of Livestock Services." World Bank Discussion Paper 163. World Bank, Washington, D.C.

Univehr L., and N. Hirschhorn. 2000. "Food Safety Issues in the Developing World." World Bank Technical Paper 469. World Bank, Washington, D.C.

Vergriette, B., and J. P. Rolland. 1994. *International Livestock Markets and Impact of Devaluation of CFA on West African Pastoral Economies*. Paris: Solagral.

Wamukoya, J. P. O., J. M. Gathuma, and E. R. Mutiga. 1997. Spontaneous Private Veterinary Practice in Kenya Since 1988. Paper presented at the First Electronic Conference, Principles for Rational Delivery of Public and Private Veterinary Services, Food and Agriculture Organization of the United Nations, January 1–20, 1997.

World Bank. 1983. *Philippines: Second Livestock Development Project: OED Project Performance Audit Report*. World Bank, Washington, D.C.

_____. 1989. Sudan: *Livestock Marketing Project*. Operation Evaluation Department final report. World Bank, Washington, D.C.

_____. 1994. Indonesia Smallholder Cattle Development. OED report. World Bank, Washington, D.C.

_____. 1996. Nigeria: "Second Livestock Development Project." OED memorandum. World Bank, Washington, D.C.

_____. 1997. *Rural Development: From Vision to Action: A Sector Strategy*. Environmentally and Socially Sustainable Development Studies and Monographs No. 12. World Bank, Washington, D.C.

_____. Forthcoming. Reaching the Rural Poor: The Rural Strategy Revisited. World Bank, Washington, D.C.